Profitable computer aided engineering

Profitable computer aided engineering

Norman W Sandland BSc; CEng; MIEE

WOODHEAD PUBLISHING LIMITED

Cambridge, England

Published by Woodhead Publishing Ltd, Abington Hall, Abington, Cambridge, CB1 6AH, England

First published 1992, Woodhead Publishing Ltd

© Woodhead Publishing Ltd

British Library Cataloguing in Publication Data
A catalogue record for this book is available from the British Library.

ISBN 1 85573 082 0

Designed by Andrew Jones (text) and Chris Feely (jacket), typeset by MFK Typesetting Limited, Hitchin and printed by St Edmundsbury Press.

Contents

Preface   xi

**1 History and background to computer
    aided engineering**   1

Brief historical overview   1
The 1960s   2
The 1970s   4
The 1980s   8
Current trends   11
The development of the supply market   12

**2 Forcing factors governing the development
    and use of computer aids**   17

Introduction   17
Skill shortages   18
Labour costs trends   19
Quality considerations   20

Legal pressures on engineering companies   23
Product/system life cycle requirements   25
Development timescales   26
Falling costs of computer hardware   27
Software developments   28
Trade union attitudes   29

3   The scope of the technology   33

Conceptual modelling and design   34
Design simulation and analysis   36
Design drawing activities   38
Other production and assembly documents   39
Process planning   42
Production planning   46
Production control   46
Materials management and control   47
On site installation and commissioning   48
Inspection and testing   49
Maintenance   51
As built documentation   52

4   Overview of the system components within the computer aided technologies   55

Computer hardware   56
Centralised computer systems   56
Engineering workstations   59
Personal computers   61
Graphics display terminals   61
Tablets and digitisers   63
Screen menus   64
Plotters   64
Networked systems   66
Software components   70
Operating system software   70
Application software tools   71
Application software modules   79

5 Justification for introducing computer aided engineering technology 81

Assessing sources of benefits 82
Specific company functions suited to computer aids 84
Shortening lead time 85
Improving consistency and quality in technical documentation 86
Conclusion 92

6 Using computer aids in conceptual design 93

Exploring external appearance 94
Design sketching and modelling 96
Viewing models with surface properties 103
Conceptual design performance—analysis and simulation 104
Sharing information and resources 105
Linking concept design data to downstream functions 106
Codified designs 110
Conceptual design of equipment packaging 113
Automotive and aerospace conceptual design 114
Conclusions 116

7 Computer aided draughting 117

Information growth in the product/system cycle 117
Applicability of computer aids 118
Drawings well suited to computer aided draughting 119
Techniques for dimensioning 122
Potential gain in productivity 127
Functional requirements of a computer aided draughting system 128
Handling the computer data 134
User groups 135
Vendor support 135
Systems response times 135
Drawing data transfer 136

8 Computer aided process planning 137

Alternative computer aided process planning
strategies   141
Variant type computer aided process planning
systems   144
Generative computer aided process planning
systems   145
Process planning language   149
Technical data specific to the company   149
Numerical control part programming   150
Part programming system requirements   151

**9 Computer aided material and production
planning and control** 159

Understanding the influence of company
priorities   160
Part numbering   162
Range of topics handled within an integrated
system   163
Parts and materials management   164
Works order processing   173
Purchase order processing   177
Sales order processing   180
Production management   185
Conclusion   195

10 Choosing a proprietary turnkey system 197

External consultants   198
Technical review of the company   199
Functional specification   203
Seeking vendor responses   205
Benchmark testing   209
Final vendor selection and contract
considerations   215

11 Planning the implementation   **219**

Equipment siting   219
Planning early system use   221
System administration   225
Parallel running   226
Back-up and error recovery   227
Extending the training and workload   231
Archiving data   231

12 Monitoring the system   **233**

The learning curve   235
Reviewing the operational use of the system   235
Monitoring response times   237
Reviewing data security procedures   238
Application software performance monitoring   238
Technology champions and antagonists   239
Middle management attitudes   240
Monitoring the administration of the system   242
Monitoring the supplier of the system   243
Decisions on expanding the use of computer
aids   245
Possible reasons for poor results   245
Conclusion   246

13 The problem areas   **249**

The supply market   250
Converting to using the technology   251
Components within the technology   253
Interfacing users with systems   255
Interfacing systems with systems   257
Perceptions relating to the technology   261
Conclusion   264

14 Future horizons 267

The limitation of obsolescence   268
Evolution of operating systems environments   269
Development in peripheral devices   269
User interface developments   270
Interpreting scanned data   271
Automated design   274
Intelligent systems   275
System integration   280
Educational trends   281

Index 282

Preface

Engineers respond with a wide range of reactions to the word 'computer'. Many practising engineers who have reached middle or senior technical management posts now recognise that computer aids are important, if not essential. They perceive that computer aids can help to ensure that products are designed, manufactured and tested to a degree of thoroughness which is needed to remain technically competitive in today's markets. Others, often in key management positions at different levels within many companies, still see the use of computers as being of rather doubtful benefit. In the experience of many companies, even today, the commercial benefits from the use of computer aids in the engineering related functions are still patchy. This mixed experience is found across the full range of company sizes. It also varies from country to country.

The purpose of this book is to discuss a wide range of computer aided techniques which are commercially available from computer system vendors. It has also been my plan to indicate to readers, where appropriate, how the technology may be extended in scope within a company. Often, it will be seen that in order to maximise the potential benefits from the technology, a wider view of computer aided engineering techniques is required. Senior management must become involved in its strategic planning if the risk of mistakes is to be minimised.

Many of the common reasons why computer aided technologies in engineering fail to live up to their potential will be identified. Frequently, the factors limiting the success of the technology relate to people, their attitudes and training as much as the quality of technology deployed or its mis-application. Just as the management and staff of companies have been required to adjust to the introduction of commercial Data Processing in the control of stocks, sales and purchase order processing, and financial and management accounts etc, they must also adjust to the use of computer aids in engineering.

My experience of working with both developers and users of the technology stretches over 20 years and many dramatic developments in the technology have happened in this time. This book concentrates on the application of the technology in small to medium sized companies with a sales turnover typically up to $100 million. It is not my intention to concentrate on companies operating in any particular engineering discipline. The content of the book will have general relevance to companies operating in the following broad fields of engineering:

— Industrial design;
— Mechanical design and manufacture;
— Hydraulic design and manufacture;
— Electrical design and manufacture;
— Electronics design and manufacture;
— Process engineering;
— Jobbing and sub-contract manufacture.

As is my nature, I have taken a pragmatic engineering view of the application of the technology throughout. It is hoped that this approach will assist engineers at various levels of management in assessing the applicability of the technologies described within the enterprises in which they work. It should also help them in planning for the introduction and monitoring of the technologies described so that the potential commercial benefits can be assessed and achieved. The capital and training investments which are implied can be substantial. The company management must therefore strive to gain the fullest benefits.

As with many technologies the use of computers as aids to engineering activities has become littered with abbreviations. The technical press and sales promotional literature often uses such abbreviations as the following:

CAD: Computer Aided Design;
CADD: Computer Aided Design and Draughting;
CAM: Computer Aided Manufacture;

CAT: Computer Aided Test;
CAPP: Computer Aided Process Planning;
CAPC: Computer Aided Production Control.

Where such abbreviations are used within this text they will first be defined in order to avoid confusion due to different uses of the same sets of initials elsewhere. As with any other relatively recently developed technology, its exponents and developers have caused some confusion by not having agreed standards at a sufficiently early stage of the development. This has done little to reduce the scepticism of many older members of the engineering fraternity.

I should like to thank both users and suppliers of the technology for their permission to include photographs of their work within this book. The following companies have provided photographs, computer generated drawings and other documents:

C & J Clark International Ltd;
CADCentre Ltd;
Computervision Corporation;
Delcam International plc;
FEGS Ltd;
Kewill Systems plc;
Racal Redac Group Ltd;
Robary Ltd;
SD-Scicon Ltd;
UK Ministry of Defence;
Warwick Evans Design;
Whessoe Computer Systems.

I should also like to pay tribute to the many engineers and software writers within this new and exciting industry, many of whom have made major contributions to the development of the technology. It has been my privilege to have known and worked with many of them, both within the supply companies and those user companies which have done so much to advance the technology over the last 25 years or so. Many of those I have known and worked with in the field have received little recognition for the outstanding progress which has been made. Some have made their pot of gold (and they deserved it) but very many have only the satisfaction of a job well done.

As many people have recognised it is not the technology itself which is important, dramatic though it can be. What is important is how it is employed and used within a company and how management and staff use its results, monitor its effectiveness and take appropriate actions.

Norman W Sandland

History and background to computer aided engineering

Brief historical overview

The early development of computer aids for the engineer started about 25 years ago in the latter half of the 1960s. Commercial data processing, aimed at the control and management of the commercial data which flows with a company had been in place to some degree in the larger companies for a decade or so earlier. However, the nature of engineering applications generally demands considerably greater computing power. Much of the early engineering related technology work was initiated within university departments in the UK (notably Cambridge University) and more widely in Western Europe and the West Coast and New England areas of the USA.

Conventional commercial data processing involves relatively simple computations with a large volume of generally large alpha-numeric interrelated records. The emphasis is on maintaining the integrity of the relationships between data records, providing good auditability of the transactions which have occurred and providing a flexible methodology for extracting and reporting data in differing combinations. The volume of alpha-numeric data in such applications is often large. The early commercial data processing applications were also invariably run as batch operations in which there was little

1

interaction by the user with the exception of loading data into the system. Also, the use of computer graphics techniques was extremely limited in these applications, using occasional graphs, charts and simple diagrams only.

Engineering application of computers generally involve a different combination of computing parameters. For many applications concerned with design simulation and analysis the volume of input parameters can be small (except for system simulations concerned with the electronics and process industries). However, the computations implied are frequently complex. They are often iterative in nature and the volume of output data is sometimes quite small. In some notable occasions, the volume of output data can be extremely large, particularly for time related simulation data or for models of large and complex systems such as electronic or process plant simulations. Similarly, large volumes of both input and output data are to be found in cases such as the three dimensional modelling of process plant, finite element stress analysis or for building other complex models etc.

The 1960s

From the outset, it was realised that an important aspect of using computers to aid engineers was the need to model engineering drawings and diagrams. Much of the communication between different functions within engineering is codified in the form of drawings and diagrams, often containing numeric and/or textual information which is directly related to the geometric entities contained within the drawing. Thus, it was realised in the mid-1960s that, if progress was to be made in using the potential of the computer for engineering applications, initial research must in part be focused on the development of computer graphics terminals and computer software to generate a rich set of graphical functions.

Storage tube graphics terminals

Early limitation of computer hardware computation speed and the relatively high cost of computer memory led to the dominance of computer graphics terminals based on storage tubes. These had been used in electronics within oscilloscope test equipment. The main advantage of the storage tube was that the graphical image generated by the engineering application software, need be generated only once by the computer as a result of any interaction by the user. It was then held electrostatically within the storage tube terminal until the terminal screen was cleared by the user and the picture redrawn (probably in some modified form). This obviated the necessity to either store the graphical image to be displayed in the computer or in terminal memory. It also eliminated the need to continuously regenerate the image as on a raster frame, similar to that which is used in domestic television.

A further advantage of the storage tube graphics terminal was the relatively high spatial resolution which could be achieved for a modest cost. Also, associated hard copy devices were available which could generate a hard copy image of what appeared on the screen within a few tens of seconds, albeit limited to A4 sized paper copies.

The main disadvantage of these early graphics terminals was that if the graphical image was changed, as a result of interaction with the user, the old part of the graphical image could not be selectively erased. The new graphical elements could only be added to the stored image until the screen was cleared and the new modified image redrawn on the screen as a result of a command from the user. Also, these storage tube terminals did not generally support colour and graphical information could not be dragged across the screen in a dynamic manner.

The combination of these two limitations often led to the graphical image on the screen becoming cluttered and confused. Parts of the information were crossed out (to indicate that a geometric element had been deleted) until the user called for the screen to be redisplayed. For large complex engineering drawings or models the time to redisplay the full screen of information could become long (sometimes several minutes).

Refresh vector graphics terminals

In parallel with the development of storage tube graphics terminals, attempts were made to develop refresh vector graphics terminals. With this terminal, individual graphical image vectors were continuously refreshed in sequence on the graphics screen. This enabled the computer generated graphical image to be rapidly updated as soon as the geometry of the graphical image was changed by the user. This approach required digital memory to store all of the graphics vectors. The graphical image memory was initially provided from within the computer's own memory but later additional memory was included within the terminal itself. The refresh vector approach clearly required less memory than that required for a full video image of a raster frame, since only the end co-ordinates need be stored for a given maximum number of vectors. Such refresh vector terminals could support good spatial resolution on the screen image. Graphical images could be generated dynamically on the screen since the image was constantly being refreshed. They were however significantly more expensive than the storage screen terminals for any given screen size and spatial resolution.

Another disadvantage of the refresh vector graphics terminal was as the number of vectors increased within an image to be displayed, so the time to refresh the screen increased and the interval between successive images being updated on the screen became longer. Due to the then limited computation speed, this produced annoying flicker on the screen for graphical images containing more than a relatively small number of vectors (a few hundred). Given that text characters and geometrical shapes such as circles and curves are each made up of a number of short linear vectors, the number of graphics vectors soon exceeded the limit when flicker became a problem.

The earliest computer aids for engineers demanded the availability of comparatively large main frame computer processors - then operating at optimum clock rates of about one megahertz (million clock impulses per second). Such computers had to be shared between a number of engineering users, using a mixture of storage tube graphics terminals and alpha-numeric VDUs. Each user was serviced by a time-sharing operating system in which the user would be allocated short bursts of central computer resource and disk storage access. Graphical information, whether in the form of conventional engineering graphs or engineering drawings and diagrams, was serviced in a plot queue which in turn was run at low priority in background mode.

The 1970s

Mini computers

By the early 1970s mini computers were being brought to the market. These were largely of American manufacture but marketed on a worldwide basis and supported with virtual memory multi-user operating systems. This combination of computing support offered a good platform upon which computer aids for engineers could expand and develop. Although initially the processor memory was somewhat limited, often under a megabyte, it allowed both software development teams and engineering users to be serviced with their own computing resource. They were capable of handling what were then becoming large software programs (several hundred kilobytes). The processor memory also allowed the engineering functions to extract themselves from the computing resources, used primarily for the accounting and commercial data processing management functions within companies. Bold engineering managers could now buy complete hardware/software packages from a highly fragmented supply market aimed at meeting their specialist needs for engineering computer aids.

By this period, mini computer based systems which were accessed via so-called 'dumb' alpha-numeric and storage tube based graphics terminals, were available at the sorts of cost which a production manager may have to pay for a new machine tool (a few tens of thousands of dollars). The central mini computer based system would be supported typically by a quarter of a megabyte of random access memory, 300 megabytes of removable disk, a drum or flat-bed plotter and a tape unit for backing up purposes. Such configurations would typically service about four or five simultaneous users who would be operating in contention for the central computing resource.

By the mid-1970s such hardware configuration supported engineering applications such as:

— Electrical schematic draughting;
— Two dimensional mechanical drawing;
— Vibration analysis;
— Printed circuit board topological design;
— Piping and instrumentation diagrams;
— Piping isometrics;
— Process simulation;
— Building and highway design;
— Numerical control part programming.

The main limitation of the technology based on main frame and mini computers, operating in contention and using so-called 'dumb' terminals was the inconsistency of system response time. The mini computer had to be time-shared between all of its users who undertook a mixture of tasks, many of which imposed a high computational load on the single processor. Initially, all computer processing associated with the display of information had to be undertaken within the same mini computer. Engineering computing applications are by their nature often highly interactive. Serious problems often occurred when a number of users were simultaneously using computer programs which made a high demand on the single central computer processor or on the disk storage system. Such contention for the processor or disk system in a multi-user time-sharing environment often leads to a wide variation in response time of the system as observed by its users.

Operating system software

At its simplest level, the software used within a computer in order to complete useful tasks, can be considered in two parts. The first is the system's operating system software. This takes care of the internal control and 'housekeeping' including file handling and control of the peripheral devices attached to the computer processor (e.g. input devices and output devices). It also allocates computer and associated resources to the multiple concurrent users.

Application software

The second is the application software which performs the tasks required to assist the user in application tasks (e.g. computer aided analysis, computer aided draughting). The operating system is provided by the original supplier of the computer hardware. The application software for engineering is written by engineering software houses who, in turn, tend to supply full hardware and software packages to the engineering user market.

A designer, draughtsman or production engineer may tolerate longer response times when the computer is asked to compute a difficult task (such as fitting a smooth surface to a series of three dimensional points in space or computing the area of a complex shape). Such users however rapidly become intolerant of slow response times for such tasks as asking the computer to find a symbol on a circuit or piping diagram at which the user points, or a line/circle on a mechanical detail drawing. In this second case the user can see the item of interest on the screen and perceives the task of locating it as simple, expecting instantaneous response. In fact the computational load in the second case may be of the same order as the first case which is perceived as being difficult.

What is more intolerable, however, to the engineering user is any significant inconsistency in response time to commands. Such variations in response time, for the same or equivalent task, tend to interfere with the flow of thought processes of the user and can lead to great frustration and errors. These early shared single processor systems often suffered from variations of response times due to changes in total computational or disk access loads on the system.

Virtual memory multi-tasking operating systems

The suites of computer programs needed to support engineering applications, whilst modular in construction, quickly became very large by the standards of conventional commercial data processing. Multi-megabyte application program suites became available as early as the mid- to late 1970s. A number of major mini computer system suppliers had recognised the need to efficiently service large programs. They had developed so-called virtual memory multi-tasking operating systems. Such an operating system environment took care of loading, unloading and executing application software modules (from disk) in segments which were small enough to fit into the available user partitions within the random access memory of the computer. This virtual memory environment allowed the application software developers to expand the scope (and hence the size) of their software systems. These developments were aimed at improving the functionality of the computer aids to meet the perceived needs of the engineering user community.

The use of virtual memory operating system environments however compounded the problem of servicing a number of users. Each user often used different application software modules which, together, formed large application software suites. The virtual memory operating system is required to swap application software in and out of the users' memory spaces in order to respond to the application commands requested. This process can involve a high frequency of disk access. The problem becomes more severe if the internal random access memory of the computer is limited and the structure of the application software is not well matched to the likely sequence of operations demanded by the user.

It was not unknown for a heavily loaded system to finish up spending most of its time swapping application software in and out of the computer's memory from its disk storage. In this situation there was little time or disk access resource left over for the computer to do any useful work. Consequently, in these circumstances the response time to the engineering users increases alarmingly and can render the system virtually redundant.

Clearly, the system developers needed to overcome these problems as the size and complexity of computer aided systems for engineers grew in complexity and size. In the first instance additional smaller computer processors were added to front-end the main processor and to handle some of the general tasks associated with communications with such devices as the user terminals and plotters.

Computer networks started to evolve where different sized computer processors were used for different engineering tasks. Large and more powerful processors were dedicated to tasks which presented a high computational load, such as finite element analysis, chemical process simulation or automatic routing design of printed circuit boards. These networks took various forms and were used for transferring both application programs and/or engineering data between different computer processors. They provided access from the user terminals to a variety of processors and application software modules.

The 1980s

By the late 1970s and early 1980s a number of major hardware developments occurred which facilitated the further development of computer aids for the engineering community. The rapid development of random access memory integrated circuits resulted in a very substantial reduction in the cost of computer memory. Memory chips became faster and the number of memory elements housed on a single chip grew exponentially. Similar developments in processor hardware chips resulted in a rapid fall in the real cost of computer hardware whilst increasing the computing power. These improvements were in terms of the numeric resolution and the computing speed of the hardware equipment.

Engineering workstations

So-called 'engineering workstations', with ever increasing amounts of random access memory, and with much cheaper, large volume disk storage units became available. This was coupled with rapid development of graphics display terminal devices in which high resolution raster technology became available together with colour. The memory and processing speed required to refresh the raster screen became inexpensive. They could be included within the terminal device itself. This enabled the main processors within the engineering workstation to be fully dedicated to running the application software of the computer aided system.

Families of such engineering workstations evolved. Different combinations of computing power, speed of processing, memory storage capacities, physical screen size, spatial resolution and number of colour options became available. These combinations resulted in flexible families of increasingly powerful workstations. Also, at this time there were rapid developments of local and wide area computer networking techniques. These ran at faster speeds than had been possible previously and allowed the large files and application programs, implicit in computer aided engineering, to pass between nodes on the network within acceptable transfer times.

The application software could reside within the workstation and the full resource of the workstation could be made available to the engineering user, ensuring both fast and consistent response to user commands. By matching the configuration of the workstation, from within the family of workstations available, with the likely software load demand within such a network, the system can be optimised for a given engineering installation.

In such networked configurations it is possible to access application software and data which is resident on any workstation node on the network from any other node. In many applications a number of users may wish to access company-wide data files (e.g. data containing company standards, component information, common project data). As a general rule it is inadvisable to have master data held within the system in duplicated form, since this presents the added difficulty of keeping all duplicates up to date and consistent with other data within the company. This was overcome by centralising both application programs and master data on large disk storage units (at least several hundred megabytes in capacity). This disk capacity is serviced, in terms of access and file maintenance operations, by a dedicated computer processor at a network node. This processor services the large bulk of the master files on the network and is known as a 'file server'. Some nodes on the network will not be expected to generate a heavy load. Perhaps it is a node used by managers merely to interrogate data, or for applications which are low in computing demand. The workstation at such a node may not require its own disk resource. Such nodes are known as disk-less nodes and by their judicious use the hardware cost of the networked system can be reduced.

Screen windowing techniques

During the 1980s both operating system and application software were further developed. Original work undertaken at Palo Alto laboratories produced a powerful approach to screen windowing techniques. These allowed the screen to be split into different areas which the application software could assign to different purposes for the convenience of the user. Multiple windows on the screen simultaneously could be used to show different subsets of application data (e.g. different parts of a drawing at different scales, or component textual data overlaid on a design drawing). These techniques could also be used to provide a hierarchy of screen menus from which the user could select commands in a structured manner. Further, windows can be used to provide 'help' text, temporarily overlaid on screen. This can assist the user in understanding either errors which the application software has detected, or in explaining the use and operation of commands available. Such advice is provided in the context of the operations which the user employs.

The wider and cheaper availability of graphics screens which constantly refreshed screen images, with optional colour, within the engineering workstations allowed greater clarity of engineering drawings, diagrams and text. These developments improved the communication between the engineering users and the computer aids which they were increasingly using. Interaction with the system increasingly occurred by means of screen menu selection, using a cursor positioned by such devices as a joystick, thumb wheel, tracker ball, stylus or puck. A graphics tablet or digitiser was also used to hold menu templates which could be customised to meet the needs of the engineering application.

Most systems came to offer a choice of man-machine interfaces in which commands could be entered through a tablet menu (with multiple templates), on screen windowed menus or cryptic keyboard alpha-numeric commands. The choice of such interaction is highly subjective and system vendors recognised that they must offer a choice to the user.

Functionality

During the 1980s more and more functionality was built into the application software which had come to permeate almost all engineering disciplines. These ranged over electronic or chemical process system simulation and design, three dimensional coloured conceptual design linked to full production detailing, three dimensional design of process plant, production engineering process planning and numerical control part programming. As more functions within engineering were supported by credible computer aids, the pressures increased for the linking of engineering functions within the computer. The volume of engineering data explodes from the initial small volume of data defining a design concept, through the engineering detailing phases to the large volumes of data required for full production, inspection and testing of a product or system.

For example, it is important that the component detail geometry data, which is produced in the drawing office on a computer aided design and draughting system, should be capable of being extracted in a form which is directly compatible with requirements in the downstream functions. These functions include for example, the production planning engineer - responsible for preparing production process plans, or numerical control part programs for driving numerically controlled machine tools. Similarly, the inspection drawings required for a mechanical component or assembly which has been designed and detailed on such a system, should become a direct by-product from the system. In the world of the electronics engineer, a computer aid can be used to capture and edit a circuit diagram of an electronics system, using standard symbols and company standard components. The connectivity information, held within the computer model of the circuit diagram, should be directly available to both a logic simulator software package and a suite of programs for aiding the topological design of the printed circuit board.

Application software modules have matured to solve real engineering problems in a co-ordinated manner. These aids assist in performing tasks across a wide range of engineering disciplines. Moreover, they use directly equivalent computer models of the documents which have been traditionally used. From these models familiar engineering documents can be accurately and consistently printed or plotted.

Current trends

The latter half of the 1980s through to the present has seen rapid advances in the cost/performance ratio of computer hardware with the advent of the 16 and the 32 bit personal computer (PC), associated networking techniques and transparent multi-tasking operating systems. Early 16 bit personal computers, supplemented with a maths co-processor chip, were comparable in power with the lower end of the mini computer ranges of the early to mid-1970s. By the latter half of the 1980s the initial cost of widely used, but less computationally demanding, computer aided engineering applications (such as two dimensional mechanical and schematic draughting) had migrated to the personal computer range. The advent of the 32 bit word length personal computer, with clock speeds of 25 to 40 MHz, increased random access memory (multi-megabyte), support for larger disk storage units (several tens of megabytes), and the availability of operating systems which could handle program segments in excess of the 640 k limitation of MS-DOS (and equivalents). This allowed a wider range of computer aids to be credibly supported. There is now a clear overlap between the capabilities of the top end of the personal computer based networked systems and those systems based on the range of engineering workstations.

However, it remains the case that the computing load presented by a wide range of computer aided engineering applications requires computer processors well beyond the capabilities of the PC. Such tasks as credible auto-routing of printed circuit board designs, finite element analysis techniques, chemical process simulation, electronic system simulation and three dimensional modelling still need to be resourced by more powerful computer processors. Such applications still demand the resources of the top end of the range of engineering workstations, mini computers and main frames.

Throughout these developments over the last 25–30 years, there has been at least one serious omission in the development strategy. This concerns the general lack of data compatibility between vendors' systems aimed at the same engineering market sector. The result is that a mechanical detail drawing generated on a computer aided draughting system supplied by system vendor A is not directly compatible with a similar draughting system supplied by vendor B. Similarly, a printed circuit board design produced by a PCB design system offered by vendor C is not directly compatible with a similar PCB system supplied by vendor D.

This situation came about because of the somewhat piecemeal development of the technology by a wide range of large and small companies spread mainly throughout the UK, Europe and North America. This problem was recognised by the mid-1980s and efforts were made to try to establish so-called neutral interface formats for computer graphics and engineering application data. As will be seen later, this approach has only partly solved the problem of passing data between dissimilar vendors' systems. The other approach has been to write dedicated conversion software modules which translate the format of data between the computer system on which it was generated and that to which it was passed. This latter approach requires that a separate translation program is required for each pair of vendors' systems. This can be both costly and time consuming in a user community where there are systems in use from a wide range of vendors.

The development of the supply market

The supply market for engineering computer aids has gone through a number of quite severe changes and rationalisations during the last 25 years. The pattern of change has been similar in Europe, North America and the Pacific Basin. These developments have in part been dictated by the evolution of hardware.

In the late 1960s and early 1970s there was a strong concentration of companies supplying engineering software to the engineering community via time-shared computer bureaux. This was largely due to the original need to support engineering computer aids using main frame computers. Many of these early applications were largely concerned with design analysis and simulation. They were associated with stress analysis, vibration analysis, electronic circuit simulation and chemical process simulation et al. Such computer bureaux companies often offered their services on a multi-national basis with technical support on a regional basis. Users accessed the technology via modems through communication lines provided within the public telephone network. They paid for access to these computer aids on a formula basis, calculated on the terminal connect hours, central processor units used and a software royalty.

The advent of mini computers and later engineering workstations and personal computers, coupled with the development of engineering applications containing a higher degree of computer graphics, resulted in a great explosion of system vendors. These new companies were largely software houses staffed by a combination of software specialists and recently practising engineers. These teams worked together to develop, sell and support a wide range of computer aids targeted at ever widening markets in the manufacturing, process and civil engineering sectors. Such companies developed rapidly from the mid-1970s onward and were largely based in the USA, UK, France, Germany and Italy.

These pioneers of the technology were characterised in two main groups. The larger companies, which were predominantly American owned, offered a wide range of computer aids covering a number of different functional areas across a number of industry sectors. Other companies developed system products aimed at particular niche functions within selected industrial sectors. These specialist computer aids included such functions as computer aided draughting systems; systems aimed at the electronics industry covering circuit diagrams, logic simulation, printed circuit board design, automatic test pattern generation; numerical control part programming systems and architectural design aids.

The suppliers largely provided so-called 'turnkey systems'. These comprised total hardware and software systems. In theory at least, the system was delivered to users, who after suitable training, 'turned the key' and started to use the computer system on their company's engineering work – gaining all the benefits which the salesman had promised. Within the turnkey system there were some facilities for customising the computer aid in order to optimise its operation for the work to be undertaken. Computer libraries of component data (both graphical representation and operational parameters) were supplied by the system vendors and by a number of service companies which began to specialise in providing add-on facilities for the major system vendors.

Generally, the turnkey system vendors used more or less standard hardware platforms upon which to build their systems. Some of the larger American vendors attempted to design and build their own hardware, in order to increase the added value of the hardware manufactured within their business. These companies generally hit significant problems and eventually reverted to using standard proprietary hardware.

By the mid-1980s the supply market was dominated by specialist niche vendors who were the originators of large software systems and by a small number of very large, largely American owned computer hardware manufacturers. The latter had negotiated the marketing rights and sometimes acquired the software development teams of the computer aided engineering software products. The last half of the 1980s saw considerable rationalisation of the major system suppliers with mergers and acquisitions. A somewhat larger concentration of the supply of larger and more complex computer aids has now passed to the control of the predominantly USA owned long established hardware manufacturers.

The situation at the smaller system end of the market (e.g. general purpose draughting, interactive PCB design) has changed very markedly with the rapid development of the power and speed of personal computers. Many of the software systems which were launched into these large market sectors in the mid-1980s failed to survive. The failures occurred often because of lack of functionality, under-capitalisation of the developers or poor marketing and support. A relatively small number of well developed application software systems aimed at general engineering have emerged with dominant market share within the industry. These systems are now sold by a wide range of sub-agents who sell turnkey networked systems into these major market sectors. They are often based on computers at the top end of the personal computer range. It has to be said however that for many (though by no means all) of these vendors, the main motivation is to sell personal computer hardware, on which they make their commission. Often the application software support leaves much to be desired from such vendors.

The last 25 years have seen great strides in the development of both computer hardware and engineering application software. The take up of the technology within industry across the western world has indeed been dramatic, particularly since the advent of the personal computer. This has allowed many small and medium sized companies to make some measure of investment in these new technology aids, if only in the area of the design drawing office. By the start of the 1990s the worldwide market for computer aids to all engineering was about $7 billion per year. Over 50% of this total was being sold into the mechanical engineering sector of manufacturing industry and about one-third into companies concerned with electrical and electronics engineering. The largest growth sector of this market was in personal computer based systems.

The USA users dominate the market place with about 50% of worldwide purchases. Within Western Europe, the UK and Germany each represents about a quarter of the market with France running about a fifth and Italy, Benelux and Scandinavia purchasing on or around 10% each.

2

Forcing factors governing the development and use of computer aids

Introduction

The factors which have combined to promote the development of computer aided technology for engineering and its use within smaller to medium sized firms will be discussed. The following topics are amongst the key factors which have had an impact on the development of the technology:

- Skill shortages;
- Labour cost trends;
- Quality considerations;
- Legal pressures on engineering companies;
- Product/system life cycle requirements;
- Development timescales;
- Falling costs of computer hardware;
- Software developments;
- Trade union attitudes.

Any emerging technology will only find its place in the market if it serves to solve the problems faced by its potential customers at a cost which will be supported by the market. In the case of computer aids for engineering

17

the growth in sales largely came about as a result of a combination of 'technology push' and 'market pull'. A number of the larger companies engaged in the automotive, aerospace, electronics, process plant and civil engineering sectors had started to make use of computer aided techniques prior to the beginning of the 1980s. However, it was not until the early 1980s that the technology began to be taken up, in any serious way, by, at first, medium sized companies (with a turnover of say less than $100 million) and later by smaller companies (with a turnover of less than say $20 million).

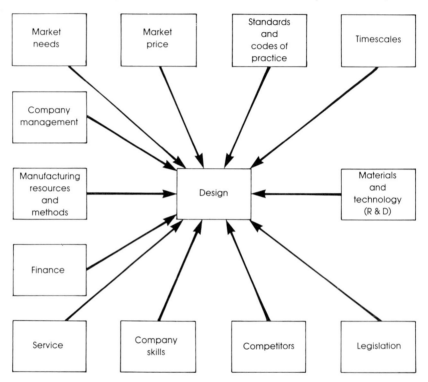

The inputs to design.

Skill shortages

The start of this major expansion period coincided with the major recession of the early 1980s which had resulted in widespread lay-offs of unskilled, semi-skilled and eventually highly skilled labour. Training of replacement staff fell sharply. In the UK, engineering training had never really recovered before the further recession in 1990 set in. Indeed many senior engineers and technical staff, serving in manufacturing industry in the early 1980s, expressed the view in the UK that if the country ever recovered from the manufacturing recession of the early 1980s it would no longer have the skill base to service a revival in its then reduced manufacturing base. In the event, to some extent, this was the case and is part of the reason why the UK has become a net importer of manufactured goods.

Whilst the situation on skill levels in most of the rest of Western Europe and the USA was less severe, the memories of the early 1980s have led to a very patchy picture on skill availability in the western world. The further recession which befell manufacturing industry in 1990 in the UK together with the demographic decline in the number of young people coming on to the labour market will continue to aggravate the skill shortages into the future.

The skills situation in the UK is even more sharply brought into focus due to the failure of the education system historically to attract young people into the sciences and engineering and the continued relative decline in the nation's manufacturing base. Even in more successful countries within Western Europe, engineering skills are not to be found in abundance at the levels required to remain fully compatible with such nations as Japan and the emerging economies of the East and South America.

The pressures have forced firms to maximise the effectiveness of trained engineering staff in order to help overcome the skill shortages. The amount of repetitive and unrewarding time consuming work, associated with preparing full documentation for a new product or system design is increasingly seen as being wasteful of the resources of trained engineering staff. In particular, the preparation of large volumes of detailed drawings, diagrams and their associated parts lists was seen as being a major area where the evolving computer based technology offered the potential to improve the performance of the technical staff. Many of these staff are found within the drawing office and production planning functions across a wide range of engineering disciplines.

Labour costs trends

Inflationary pressures across the western world during the 1970s and early 1980s have led to increasing labour costs within manufacturing industry. In the UK, engineering labour costs have been traditionally lower than its major competitors in Western Europe. This has been coupled with lower capital investment in manufacturing plant and historically much higher interest rates.

In many industrial sectors of manufacturing industry, the labour costs in the development of a full set of documentation for a new product or a custom designed system, such as a petro-chemical plant or a communication network, have increasingly become higher. Within the older industrialised nations the trend has been for their engineering markets to be at the higher technology end of the product and system range. The emerging industrial nations have tended to concentrate on higher volume, lower technology products where the lower technology base and labour costs have combined to increase their share in world markets.

The combination of inflationary trends in labour costs and the need for western countries to concentrate on the higher technology end of the spectrum of products and systems are key factors in the development of computer aids for engineers. Higher technology designs often contain proportionately a higher content of engineering design and continue to force up the labour cost content of the investment needed at each cycle of product or system development.

A key need for any computer aid applied to a given engineering function is to reduce the labour cost for that function. If this cannot be achieved, the objective must be to reduce the labour cost for the engineering related functions which are downstream of the function to which the computer aid is applied. For example, more rigorous or extensive computer based design analysis or simulation may reduce the need for later prototypes in the development cycle. Alternatively, the preparation of more comprehensive and dimensionally validated mechanical detail drawings may result in less errors and consequent scrap in the assembly and inspection processes.

Quality considerations

There is a continuing need to enhance and develop the engineering control mechanisms required to improve quality and reliability of new products and systems. It has been recognised for many years that in order to achieve good quality in design and manufacture it is necessary to develop the plans required to achieve such quality and then to monitor quality performance against the plan. In order to facilitate quality within engineering processes it is essential that engineering data is evolved in a rigorously structured manner within the design and production cycle. Moreover, in most engineering processes this data is generated in an incremental way from a variety of engineering disciplines, often involving a mutual dependence between these disciplines.

The development of a quality plan must take account of the disciplines involved and the sequence in which a product or system design is undertaken from its inception through to final inspection and/or commissioning. Clearly more rigorous analysis, simulation and testing are key factors in improving overall quality. Also, sharing of product or system data between disciplines as the engineering information evolves is a key element in maximising potential for good quality design and manufacture. The more advanced the product or system to be designed, the more need there is for its associated data to be shared between key members of the team of engineers and associated technical staff at the earliest possible time.

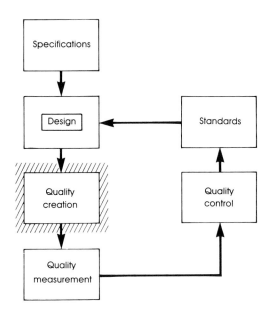

Building quality into the design process.

It is also crucial that each member of the design, development, production, inspection and test team are all working to a common set of documents which relate to the product or system. Well ordered control of the configuration of the product and the maintenance of a clearly auditable sequence of development documentation are essential if good management of quality is to be achieved. Even a modest product or system developed within manufacturing industry will involve hundreds of documents in the form of drawings, diagrams, specifications, parts and material lists and other documents. If quality control is to be maintained it is essential that these documents are kept up to date and communicated to appropriate technical staff within the engineering functions of the company.

The use of computer based technology can be advantageously employed to structure much of this data and maintain rigorous control over the compatibility between associated data.

Company standards

Many companies employ company standards, whether by way of methods of design, the use of standard components, or methods of manufacture and test. The quality function is concerned both with optimising these company standards and ensuring that they are used where appropriate in new designs. Such standardisation can be a major factor in maintaining control over quality in the engineering functions. However, many firms experience difficulty in constraining some of their engineering staff to work within the company standards which have been developed, often by trial and error over a number of years. This is often a problem with younger engineering and technician staff. Whilst it is desirable for less experienced or younger staff to use such standards, it is also important not to stunt the imagination or creative drive of such staff - a sensible compromise must be struck. Much will depend on the nature of the company and the products, systems and services which it supplies.

Computer aided techniques can be employed, making it easier for company standards to be used. For example, company standard components can be set up within computer libraries or standard tools can be made available on a tool library (within a numerical control part programming computer aid). Similarly, standard sub-assemblies within a range of products can be incorporated within a computer aided design draughting system, such that it is much more convenient for the young designer to explore and possibly use a standard sub-assembly than to set about creating yet another variant. Whilst it is often not the purpose of the computer to enforce standardisation, it can make users aware of what standard methods are used within the company. It also enables users to make an accurate assessment of whether the standards are appropriate to their current design.

All engineers must trade off the time and cost of more and more rigorous analysis and examination of alternative designs in order to improve and optimise performance, quality, reliability and cost. Indeed this iterative and sometimes frustrating activity is the very essence of the engineering process. In the days when a pound sterling was worth four US dollars, the engineer was often defined as one who could do for a dollar what any old fool could do for a quid!

The designer's compromise.

Exploring more design alternatives by using computer aids can have a major impact on both the final quality and cost of the engineering solution which is obtained. When this effect is integrated over major product development such as those found within the aerospace or automotive sectors, the result can be the reduction of the number of prototype products which have to be made whilst ensuring no degradation in final performance or quality. For small and medium sized companies, the cost and timescale savings in being correct first time can equally represent the difference between commercial success or failure.

Legal pressures on engineering companies

There have been increasing legal obligations applied to engineering companies engaged in the development and marketing of products. Similarly, companies concerned with the design of systems for use within a wide range of processes have needed to be aware of the legal consequences in the event of their systems failing. In recent years the rate of introduction of new legislation in the USA and Western Europe concerned with the safety and performance of both products and systems has accelerated. The emergence of Eastern Europe into the free markets of the world require that their manufacturing companies need to conform to international legislation.

There is now a body of legislation to ensure that good design and manufacturing practices are followed to protect the health and safety of staff. Further legislation covers the need for new products and systems to be fit for the purpose of the end-users. For a number of specialist areas of engineering design and manufacture, extensive codes of practice have evolved to minimise the risks of failures and accidents. For example, areas of design such as pressure vessel design, nuclear engineering, electrical wiring, aerospace engineering and structural design are now extensively covered by a range of national and international codes of practice which have standing in law.

There has been increased pressure within the European Community nation states for national standards and codes of practice within all of the major sectors of engineering to come closer in both scope and detail. This extra urgency has come about as a result of the 1992 single European market.

It is now essential for all engineers and technical staff who authorise design documentation or supervise manufacturing and testing processes, to take full regard of their legal responsibilities and of appropriate codes of practice. Many experienced engineers will be aware of the detail of the appropriate legislation and codes of practice. However other members of design teams will also need to conduct their tasks in accordance with legislative constraints and codes of practice. It is also necessary in international trading to take account of variations in design constraints between the home nation and that of the client company.

There is a strong requirement within much of the current legislation to ensure that sufficiently rigorous design analysis and testing is undertaken, commensurate with the possible use and subsequent damage which could be caused in the event of subsequent failure of the product or system. It is often essential that engineering documentation produced for the full design, production, test and maintenance cycle is adequate for the product or system. Where appropriate the documentation should be held in a structured manner. It should also be capable of being audited by suitably qualified engineering staff.

The use of computer aids

Computer aids have been developed in order to assist engineers in conforming to recognised national standards, see later. At the simplest level, most widely used computer aided design and draughting systems have enshrined within them an appropriate drawing standard in respect of standard symbols, special line types (e.g. centre lines for mechanical drawings or bus lines for logic diagrams) and dimensioning conventions. Similarly, design analysis software packages exist which perform design analysis or synthesis in conformity with published codes of practice (pressure vessel design, structural analysis, clothing design, electrical machinery design).

Another aspect in which designs undertaken using modern computer based technology can facilitate the objective of being fit for purpose concerns the ability to extract numerically held manufacturing and inspection drawings directly from design geometry. Techniques such as computer aided animation and thermal analysis techniques can also allow the engineer to explore the precise performance of mechanical components and moving mechanisms over the full range of operating environments. Finite element analysis techniques allow more detailed considerations of possible failure modes. Where appropriate, such techniques can be applied in order to demonstrate both a commensurate level of design responsibility and to improve the quality and cost performance of the design.

Computer based drawing and document management systems can now be used to track and monitor the status and relationships between technical documents. They can facilitate the authorisation and subsequent release and monitoring of engineering documents. The impact of design changes on the rest of the documentation set for a product or system can thus be more readily monitored and the integrity of the engineering data thus preserved. All of which tend to promote good engineering practice within a company and should go some way to ensuring that engineering functions are carried out in a manner compatible with modern legislation.

Product/system life cycle requirements

As competition increases in many market sectors, there is a constant need to review both the functionality and cost of production within a company. In many industries product life cycles have tended to reduce over the last 25 years or so. This has been due, in part, to the rapid development of such major engineering nations as Japan, Germany and some developing industrial nations. Technical innovation, improved functionality, enhanced reliability and cost performance of a wide range of products developed by these major industrial nations increasingly force companies to embark on new product developments.

In a number of important manufacturing sectors there is a strong fashion content. In such industries (e.g. the shoe, clothing, consumer electronics, wall covering and textile sectors) it is essential that companies can respond to fashion trends which significantly change in a short period of time and to bring to fruition replacement products which will remain competitive in the market place.

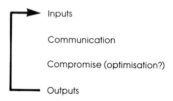

The design process.

A classic example of the modern need to meet such pressures is seen in the shoe or garment industry sector. Fashion styles for the spring fashions are dictated by the major shows which take place within the main fashion centres in Italy, UK and France in the autumn and early winter of the previous year. New products which relate to these new fashion trends must be available in a wide range of sizes, patterns and materials before the first sunny day of the following spring. Failure to generate the new designs and produce such goods at the required quality and quantity within this timescale can result in major losses for the manufacturers of fashion goods. As much as 50% of their potential market can be lost, even if they are only one to two weeks late in getting the required goods to the high street.

Development timescales

As product and system life cycles have tended to reduce, so has the pressure to reduce the development timescales for the large majority of new products and systems. Companies in much of the manufacturing industry need to bring their offerings to the market more quickly in order to meet the increasing pressures of their competitors. A shorter development timescale, providing the design is undertaken in a rigorous and professional manner, can reduce the cost of development and will often improve the cash flow potential for the company.

As the market for many products and systems becomes more sophisticated, the pressure for partly customised products and systems increases. This is sometimes achieved by selecting standard functional options and configuring these into a solution which is specific for the customer. In other cases aspects of the product or system are required to be designed, manufactured, inspected and tested on a fully bespoke basis. Here again the market advantage in being able to provide a part or full customisation service within as short as possible timescale are clear.

A structured computer based approach to combining optional standard features within a product or system design can be shown to offer advantages over traditional manually based 'pick and mix' methods. It is however essential, when using such computer based systems in a 'pick and mix' environment, to invest in setting up the information relating to options within the company specific data structures applied to the computer aided application. When these product options have been captured into the computer aided system they can be rapidly recalled, explored and incorporated within the bespoke design. This approach can be taken further within the computer. The design rules required to control the legitimate use of standard options in acceptable combinations can be built into the computer aid in order to facilitate quick, accurate and validated development and documentation of bespoke product data.

Where bespoke aspects of the product or system must rely on the new development of parts of the product, the reduction of timescales needed relies mainly on the ability to examine and document design alternatives in a shorter time. Often this can be achieved as a result of improved and more rapid facilities for changing and editing information when using computer based techniques. Also the possibility of being more correct first time is important in these cases.

Falling costs of computer hardware

Whilst the pressures highlighted within this chapter have grown, the real cost of computer hardware required to support computer aids for engineers have continued to tumble. The real cost, after allowing for inflation has regularly fallen over the last 25 years. The rates of decline in hardware cost has of course varied but, on average, the cost of computing power has roughly halved every three or four years. Alternatively, the amount of computing power offered for a given price has roughly doubled every three or four years.

These dramatic changes in computer hardware price/performance have been possible because of the rapid developments which have taken place in microelectronics components. The volume of random access computer memory available within a single integrated circuit has grown geometrically. The speed of operation of affordable computer processors has also seen increases from about one million cycles per second to 35 to 40 million cycles per second in the last ten years. The number of bits which can be processed at the same time within microprocessor integrated circuits has increased from 8 to 32 over roughly the same period.

These developments in central processors have been accompanied by parallel developments in such peripheral devices as graphics display terminals, mass storage disk units, line printers and graphics plotters. (The price/performance advances for some of these devices has not been so dramatic as for processor technology.) The advances made in reducing the cost of hardware configurations needed to support computer aids for engineers have been most marked particularly in respect of the memory and computing speed which have become available at an affordable cost.

Software developments

If one takes the long term view, software components have developed more slowly than hardware. The generation of software is highly labour intensive and it is essential to eliminate errors in the commands and structures before releasing the package for use as an aid to engineers. For most engineering applications the size of the software needed to perform the complex computing tasks implied is large by any standard. For such tasks as analysis, simulation, draughting and three dimensional modelling, each application requires hundreds of thousands of programming code lines which must be supported, maintained and further developed. Even a modestly complex application such as basic two dimensional draughting consumes several megabytes of memory when compiled into executable code.

In order to support the extensive effort to develop the required application software, it is essential to develop modular software tools to address the various aspects of the software system. The trends have been to build application software around such modules as relational database systems, language processing software tools, interactive graphics routines, windowing routines etc. These software tools have been developed to be re-usable within a range of engineering applications. Currently, further tools are emerging aimed at capturing, modelling and using engineering knowledge. These tools are supporting the introduction of so-called 'artificial intelligence' into engineering computer aids.

The large size of the software systems needed requires that the software is engineered on a modular basis in order to facilitate ease of software maintenance and trouble-shooting. The trend towards adopting an engineering design approach for software has been essential in achieving the prerequisite software quality standards. So-called 'software engineering' strategies have evolved to meet these challenges.

There has been a strong emphasis in providing the system user with a variety of interfaces with the application software. It is recognised that the engineering user wishes to use engineering language to interface with the computer. There has been much emphasis in enabling application software to use structured pull-down screen or tablet command menus. Graphical icons for use with such menus have also been developed to facilitate ease of recognition and use of commands. However, some users still prefer to use cryptic keyboard commands consisting of one or two key strokes.

Trade union attitudes

Much has been written about the failure of the trade union movement to adjust to modern methods within industry, particularly within the UK. However in the case of computer aided engineering techniques, there has been considerable pressure from the workforce for companies to adopt the new technology in order to enrich the tools available to undertake both innovative design or manufacturing and routine tasks such as draughting.

The unions have recognised that in order to achieve compétitive products which help to protect and increase jobs, the introduction of appropriate modern methods and techniques is essential. Technical staff at all levels have been exposed to the potential of these technologies through attending exhibitions and through the technical and trade press. The skills which their members acquire in using new computer aids are seen as important in negotiating terms and conditions within the place of work and to open up new career opportunities.

Whilst one of the objectives of this technology is undoubtedly to improve basic productivity that could be seen by the unions as a threat to jobs, on balance the unions have taken the view that the introduction of such aids is both inevitable and welcomed. Many companies have negotiated special labour rates for staff who extensively use computer aided techniques. In other cases, the technology has been introduced without changes to the labour rates of the operators of the equipment beyond the normal facilities of grading within staff.

Overall there has been a good attitude shown by the trade union movement in the matter of the introduction of computer aids for the engineering functions. It is however important that operational staff play a full and constructive part in selecting and planning for the introduction of the technology. There is something of an attitude of 'me too' amongst technical staff, as they see the use of computer aids within other companies and their competitors. In some cases this has created a 'bottom up' pressure for the introduction of the technology into their own companies. This 'me too' desire is often shared by senior managers within firms and leads to a 'top down' pressure to introduce the technology.

In many small and medium sized companies, these pressures, together with the lack of specialist computer expertise and the shortage of time to explore the technology in full, can lead to poor planning and monitoring of the technology with attendant disillusionment and costly mistakes.

Many mistakes were made in introducing conventional data processing technology into the commercial operations of companies in the 1960s and 1970s. Often the operational engineering staff became cynical of computers when they saw the results of these badly implemented commercial data processing systems which caused many frustrations at the operational levels of the company. With these commercial data processing systems, particularly when they were ill thought through or badly designed, the engineering and technical staff saw little output which they perceived as beneficial. Many such staff are now in intermediate and senior engineering management positions and may have a somewhat jaundiced view of computer based systems founded on their earlier experience.

It is often helpful to identify staff with such a view and to include them in the team for assessing the introduction of computer aids to the engineering functions. Such a member of the team serves the purpose of acting as a 'devil's advocate' in the planning and decision making process. It also ensures such potentially disruptive influences are exposed to the technology at an early stage of exploring and planning for its introduction. Their early involvement may win them around to the support of the technology and help to 'sell' the concepts to other sceptical staff within the organisation.

The following chapters will serve to assist those involved in exploring, planning and monitoring the use of computer aids for the engineering functions of companies. A well considered introduction and monitoring of the technology will maximise the potential benefits of the technology. It will help to avoid the many pitfalls which await the unwary section leader, manager or director of companies embarking on an upgrade to the technology base of the engineering related functions within a firm.

3

The scope of the technology

The full scope and structure of these technologies is often confused within the minds of practising engineers. Within this chapter the scope will be outlined in terms of the design and manufacturing related processes found within engineering companies. Aspects of engineering design and manufacturing cycle will be highlighted in order that the reader can form a view of the potential starting point for introducing the technology or where next to extend the use of computer based aids. The following aspects will be discussed:

- Conceptual modelling and design;
- Design simulation and analysis;
- Design drawing activities;
- Product/system production and assembly documentation;
- Process planning;
- Production planning;
- Production control;
- Materials management and control;
- On-site installation and commissioning (where appropriate);
- Inspection and testing;
- Maintenance;

- As built documentation (where appropriate).

Within this discussion of the scope, reference will be made to the impact of computer aids on the engineering activities listed above. This discussion will touch upon a range of manufacturing and process industry sectors which form the main focus within this book.

Conceptual modelling and design

Conceptual design can take many forms, depending on the nature of the design activity and the eventual design objective. The nature of any appropriate computer aid will depend on the engineering discipline under consideration. For example, different aids will assist a user if the design objective is a new product or artefact, a new component or sub-assembly or a new system of some sort. In the case of a product, the design engineer will typically start with an outline functional specification which has been prepared as a result of extensive market research and assessment. A wide range of constraints will have been established in order to position the product in the market place. The design staff will have some idea about the likely production volumes and target factory unit cost which should apply to the product.

Often the external appearance of the product is important, not only when aimed at the consumer market but also for products aimed at the professional or industrial market. The services of an industrial designer are frequently used, particularly for high volume products. For the design of a new product, traditionally the process starts with a series of two and three dimensional sketches of the outline of the product. These are used to explore alternative geometric designs. Often the conceptual designer will embellish these sketches with colour washes and provide some indication of the external surface finish.

In some cases three dimensional models are constructed, either from solid materials such as wood, plaster and wax or by assembling flat card components. Such models, together with design sketches, are used to convey the outline details of the product to other functions within the company and other interested parties.

As was discussed earlier, some product designs are strongly influenced by trends in fashion and the preparation of the conceptual design is a crucial step in meeting market needs. In some cases it is required to design a range of products rather than an individual design (e.g. in the footwear, clothing industries etc). In these cases it is often necessary to explore how the design will appear at different sizes and scales. (What may seem to be an attractive

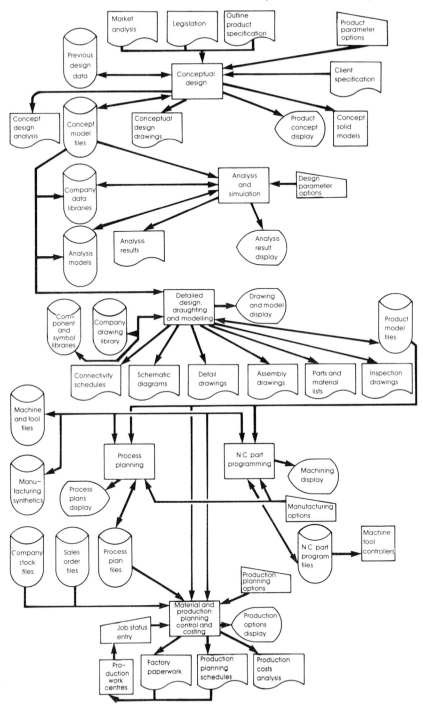

The scope of computer aids for engineers and technical staff.

concept for a shoe design at a size 7 may be quite inappropriate at larger or smaller sizes within the range.)

Many companies are concerned with the design and manufacture of sub-assemblies or components for use within a larger product which is produced often by larger companies. This is a dominant aspect of industries such as the automotive, aerospace, petrochemical and electrical power generation sectors.

It is essential that a new sub-assembly, and the boundary geometry and interfacing mechanisms of it, is fully compatible with the major product into which it will be built. This compatibility requirement forms major constraints on the conceptual design of the new sub-assembly or component. Much effort has traditionally been expended in ensuring for example, that the boundary geometry of a sun roof pressing remains smoothly within the roof line and that the sun roof fits snugly into the aperture provided in the roof pressing shell.

The conceptual design of such assemblies as mechanisms requires consideration of the relative positions of its components in various key positions of the movement of the mechanism. Exploration of different forms of mechanism design is needed. Ergonomic considerations often dictate the design concepts which need to be explored and this again can lead to repetitive exploration and 'what if' considerations.

In the case of a company which designs systems rather than products and artefacts, the conceptual design process will typically involve identifying the main subsystems within the system, together with the mechanical, electrical or hydraulic inputs and outputs and the main inter-connections between the subsystems. The conceptual design is likely to start at the block diagram level. The purpose and outline transfer function to be performed within each block will be identified in either algebraic or Boolean logic form.

Computer aids are now widely available for each of the aspects of conceptual design which have been identified above and will be discussed in detail later.

Design simulation and analysis

After the design concept has been agreed, the detail design and development can proceed. Often some form of simulation and analysis is the starting point of this design process. Most engineers perceive that the computer finds its natural place in these often repetitive and iterative processes. The interfaces between the conceptual design phase and that of detailed analysis are often not clear since in a number of cases some outline analysis or simulation must be conducted as part of the process of proving that a design concept is practical.

The range of analysis requirements is extremely wide. In the mechanical and allied fields of engineering, analysis may cover a wide range of stress, vibration, thermal and mechanism analysis. For the hydraulic engineering areas, pressure and flow analysis are crucial and in control applications, control simulation techniques are required. Within the process industries, process simulation at both the static and transient phases are essential followed by consideration of pressures, temperatures, flows, volumes and instrumentation and control analysis and simulation. In the fields of electrical and electronics engineering several types of analysis and simulation techniques are used for networks of various kinds: communications, analog circuit analysis, digital circuit simulation and analysis, control circuit simulation and analysis, instrumentation simulation, transformer and rotating machine simulation and analysis, switchgear and fault level analysis.

For certain sensitive areas of design it is necessary to embark on finite element techniques for exploring stresses, magnetic or electric fields. This involves the process of representing the volumes of material by breaking it down into a mesh upon which the analysis is performed in a highly iterative and computation intensive process.

It is within this analysis phase that the various national and international design codes of practice become important. In the process industry, heavy electrical, nuclear and aerospace sectors, many aspects of the design must be undertaken within the rules and constraints of the codes of practice which are relevant to the application. Examples include such cases as pressure vessel design, heat exchangers, conveyor system design, electrical distribution design and defence system components.

Here, the take up of computer aids is somewhat patchy. The areas of greatest dominance are concerned with finite element techniques, chemical process simulation, control system design, logic simulation, stress and vibration analysis. In these areas, appropriate analysis and simulation aids, sometimes embracing the widely used design codes of practice have evolved. Other areas of design analysis and simulation where computer aided techniques are finding favour include mechanism simulation and ergonomic design consideration.

Up to this stage of the design and manufacturing cycle the volumes of data to be handled are generally modest, except perhaps for process simulation and analysis and finite element analysis. However, as the design process passes into the main drawing office phase the volume of detailed engineering information starts to explode. Also, the general nature of the engineering documents changes. At the analysis stage the documentation is predominantly textual with often a relatively small volume of graphical performance data. The output from the drawing offices is characterised by being predominantly graphical in nature but accompanied by schedules, lists and specifications.

Design drawing activities

Even the most modest product or system design can involve the production of hundreds of drawings consisting of sketches, detail mechanical drawings, sub-assembly and full assembly drawings, circuit and logic diagrams, timing diagrams and cabling drawings. The large majority of these drawings can be considered to be one of three generic types.

Two dimensional detail or assembly drawings

This type is typically drawn to a known scale, is dimensionally significant and drawn to an internationally recognised set of drawing standard conventions.

Three dimensional drawings

This can range from sketches, through dimensioned isometric or perspectives to fully worked up scaled, coloured and textured visualisation of the component, sub-assembly or more likely, product.

Drawings with no dimensional significance

These drawings are produced to record a system based aspect of a design. Examples of this third kind of drawing would include circuit or logic diagrams, control and instrumentation diagrams. In these drawings or diagrams, symbols are typically used to represent components of various types and complexity (anything from an electrical resistor to a pump or motor). The inter-connections between the components by way of electrical or piping links are shown by lines connecting the components in various ways. The key point to note about this third type is that there is normally no dimensional significance in either the size or scale of the symbols or the length or thickness of the lines of various styles which connect them together. The drawing is purely a logical statement of the inter-relationship between the components, represented by the symbols, in terms of the links made between them.

The largest investments have been made in computer aids to support the large volumes of these three types of drawing. Predominantly, the take up and use has been associated with the production, editing and maintenance of the first and last of these types of drawings.

Mechanical component designs are toleranced and the assembly drawings are typically drawn using the geometric and dimensional information held on the final component detail drawings in order to validate the geometry and tolerances of the component drawings. Bought out components are uniquely identified with cross-references to their specification information. This stage of the design process involves structuring the design information and maintaining the integrity of the relationships between a wide range of documents.

Other production and assembly documents

The drawing output is complemented by a full set of associated documents such as parts and materials lists and schedules, connectivity schedules, inspection drawings and test schedules to form the full set of manufacturing, inspection and test documentation required to produce the prototype of the product or system. This full set of manufacturing documentation will take a variety of forms depending on the kind of product or system which is being developed.

Much of the information in these additional documents is either extracted or derived from the drawings which define the product or system. For example, connectivity schedules are derived from such drawings as circuit diagrams, logic diagrams, control ladder diagrams, piping and instrumentation diagrams. Similarly, a parts list which is associated with an electrical/electronic schematic diagram will be prepared with reference to component information held on the circuit or logic diagram to which it refers. Also, a parts and materials list which relates to a mechanical assembly drawing needs to cross-refer to the assembly drawing and to uniquely identify each mechanical component, sub-assembly or material used to build the product or sub-assembly. Further, mechanical inspection drawings and templates are a subset of the geometry which has already been drawn on the detail and assembly drawings in the design process.

The preparation of these supplementary documents is often a cause of error and omission. Connectivity links get missed off the schedule, components get left off the parts list or the quantities specified are incorrect. Inspection gauges and templates do not conform to geometry contained on the manufacturing drawing. Such errors and omissions are not picked up until a design audit or, even worse, until prototype manufacture and test.

The potential of computer based systems to extract subsets of data from the design information which is already held in data files (if the design has been undertaken using computer methods) has been exploited. For example, components on the electrical schematic can be automatically summed. By cross-reference to a file containing relevant information on each part, the formatted parts list can be generated automatically. The outline profile geometry held within the computer as one or more of the sets of bounded geometry within the detail drawing file of a mechanical component can be extracted in order to generate an inspection drawing or template. Scrap views at enlarged scale can be simply generated by expanding small areas (selected by screen windowing techniques) of a mechanical drawing held within a computer generated drawing file.

The computer aids to facilitate the production of these supplementary documents are now well developed. Since data can be extracted or derived from the same design source within the computer the potential for reducing clerical errors is considerable. Computer aids for deriving data for these

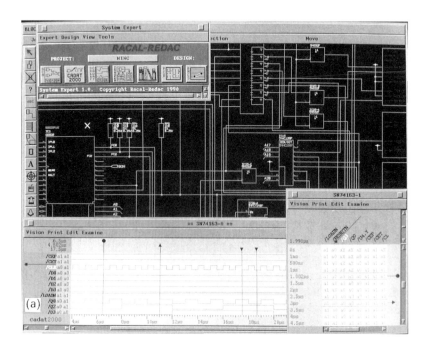

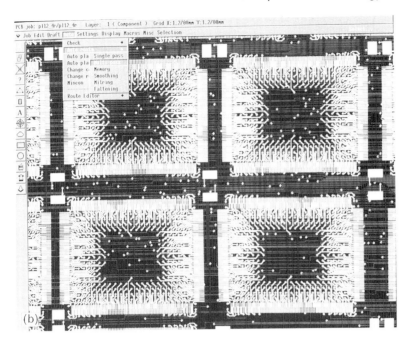

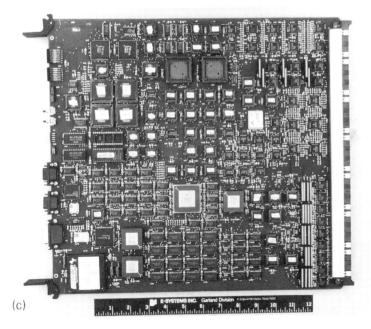

Printed circuit board design: a) Circuit schematic diagram with simulation wave-forms displayed; b) Computer generated high density tracking artwork design; c) Assembled printed circuit board (courtesy of Racal Redac Group Ltd).

supplementary documents is contained within software systems aimed at particular engineering sectors. Their use is still quite patchy and varies both within companies and within the engineering sectors. The more obvious examples such as generating parts lists for electrical/electronic/piping schematics are widely used. Extraction of connectivity information in order to link the application of computer aided schematic draughting with computer aids for circuit simulation, printed circuit board design and test pattern generation is ubiquitous. However, the automatic extraction of parts and materials lists from ballooned mechanical assembly drawings held within the computer is far less frequently undertaken.

The likely reason for this difference between the practice in say, parts lists for logic diagrams and parts lists for a mechanical assembly drawing, is interesting. In the first case, the circuit schematic has little value until each component and used element within the component level logic diagram are fully identified as to type or part number. Thus the logic diagram is normally fully annotated by the design drawing office with component type labels. In the case of the ballooned mechanical assembly drawing, full and rigorous annotation of each and every part and component with its part number or some other unique identification is less often undertaken. In this latter case the purpose of the assembly drawing is largely to aid assembly and each instance or occurrence of a geometric feature is not always drawn since multiple occurrences of a feature can be covered by manufacturing notes. Thus, in the latter case the full data required to automatically extract the rigorous parts list is not always held explicitly within the computer model of the drawing. To overcome this it would be necessary to either fully document the assembly within the drawing structure or to have some method of uniquely extracting information on parts from the free text which is used within the manufacturing notes. Hence, the pragmatic mechanical designer still often prefers to prepare parts and materials lists using manual methods.

Process planning

As the design information passes from the design function to the production planning function, there is further scope for deploying the use of computer aids. It is necessary to prepare process planning sheets which determine the sequence of operations required to manufacture or fabricate components, and to assemble sub- and final assemblies. The process planning task sometimes involves preparing further drawings which show the component or sub-assembly at each key stage of manufacture, fabrication or assembly.

Select
method
materials
tooling

Calculate
times

Document
results

The process planner's activities.

Consistency

Documentation Control

Problems with manual process planning.

The process planning stage involves using synthetic times for each type of operation which is used within the manufacturing process. These synthetic setting up and unit operation times must relate to the capital equipment, tooling, jigs and skills available to the company. They must take account of such matters as the materials to be used and the accuracy and quality constraints relating to the article to be manufactured.

Sometimes process plans are derived by modifying plans generated for earlier jobs, products, sub-assemblies or components. It is of course necessary to be able to identify a process plan previously prepared which is appropriate to the current job. This often relies on the memories of process planners or on some formal numbering system which classifies the type of job to which the process plan refers.

A special case of process planning concerns the use of numerically controlled machines, typically: lathes, milling machines, punching and nibbling machines. Here the task is to produce a program of instructions (the part program) which drives the numerically controlled machine tool in a sequence which performs the manufacturing operations in an optimum sequence. Again, appropriate part programs which have been previously generated can sometimes be used as the basis for producing the new one. The sequence of operations can be entered manually into the machine tool controller and then recorded on to magnetic media. Alternatively, they can be prepared off-line and fed to the numerically controlled machine tool either in the form of magnetic or paper tape, or downloaded directly from a central computer into the machine tool controller's memory.

Process planning, and particularly numerical control part programming, is a major area for the use of computer aids. The storage of company-specific standard data such as synthetic times, production of stage drawings and efficient and rapid recall of earlier process plans are all areas where a computer based approach offers considerable potential. Whilst computer aids to process planning are increasingly used, they offer no substitute for a company rigorously compiling accurate synthetic times which are relevant to the equipment available and conditions pertaining within the company.

Numerical machine part programming aids are extensively used and are indicated where a company has a range of different numerically controlled machine tools. There is benefit to be gained by centralising the part programming process on to a dedicated part programming system upon which new part programs can be developed off-line from the individual numerically controlled machine tools.

Production planning

Within the manufacturing function of a company, a major task concerns the planning and control of the production process. The production planning function is concerned with considering four main questions:

- What is to be manufactured? - This is defined by the set of product and system documents which define such things as geometry, connectivity, materials and finish;
- How many are to be made? - This is defined by parts lists, schedules, sales order, build-to-stock demands;
- How are things to be made? - This question relates not only to the process plan which was discussed earlier but also the available resources such as machine tools, assembly facilities and skills;
- When are the items to be made? - This consideration must consider the availability of materials, delivery targets, resource availability etc.

In order to optimise the production planning process compromises must be struck. The company has to set priorities upon which it wishes to plan. For example, what are the relative priorities between such objectives as, meeting delivery times, minimising stock holdings and work in progress, maximising machine utilisation and minimising overtime working. There are clear incompatibilities between these commercial objectives for the company. The management of the company must face the issues raised by these incompatibilities and establish clear priorities to which the production planning function must work.

Implicit in the production planning process is the need to explore different production strategies and to try a number of 'what if' production plans. In most companies, the production planning function must also make provision for unscheduled work which must be undertaken. This unscheduled work may be due to urgent orders, rework due to errors or failures under test or repair work. The company must be quickly aware of the impact of this unscheduled production work. Again this gives rise to further 'what if' exploration within the planning function.

Production control

The production control function is responsible for ensuring that production tasks are carried out within the various work centres in accordance with the production plan. It needs to control the flow of work through the various work centres and ensure that appropriate materials, documentation, tools, jigs, skills etc are available in a timely manner. In order to exercise this control and monitoring function it is necessary to issue and monitor appropriate factory documentation such as job tickets, material requisitions and routing cards.

The whole area of production planning and control is characterised with frequent examinations of alternative strategies, collation and extension of data extracted from works orders and production documentation and the need for rapid monitoring of the current status of work in progress. It has proved to be an extremely fertile area in which to deploy computer based technology. Computer aided production planning and control is widely used within most large companies and is increasingly being deployed within the medium sized companies.

Many smaller companies find difficulty in the area of planning and controlling their production as they grow beyond the stage where production can be managed on a day to day basis largely by word of mouth. The introduction of formal factory paperwork and planning charts serves to introduce the more formal production disciplines which are invariably required as the company grows. Such disciplines and a more formal approach to production planning and control are essential prior to the introduction of computer aids within this area.

Materials management and control

A natural extension of production planning and control is the associated discipline of materials management and control. A crucial factor governing the profitability of any enterprise is the effective control of parts and materials stocks. The capital tied up in stocks is a major consideration and, during periods of high interest rates, can make the difference between business failure and survival. The ideal objective is to ensure that bought in parts, materials and manufactured components and sub-assemblies become available on site shortly before they are required to meet the production schedule. In order to achieve this objective, it is necessary to have up to date and accurate knowledge of current stocks together with reliable information on the lead time for the supply of replacement parts and materials. This demands effective stock movement recording and tight stock control strategies. In many production tasks the availability of parts and materials from their suppliers will vary from item to item and also from time to time. The lead time of manufactured components and sub-assemblies is rather more in the control of the company.

From the production plan, whether this is based on building-to-stock or building-to-customer order, the quantity and timing of supply of parts and materials can be calculated. The prudent company will evolve a stocking policy which makes some provision for unscheduled work. It will recognise that the stock value is largely determined by a minority of the more expensive stock items. As a general rule, for many manufacturing companies, often about three-quarters of the stock value is represented by about only one-fifth of the stock items. Some bought in or manufactured parts will be identified as long lead time items and the planning for their acquisition is crucial if production plans are not to be ruined.

Here again, the use of computer aids for full planning management and control of parts and materials is becoming more widespread as a natural adjunct to computer based production planning and control. Computer based stock control, purchase and sales order processing systems have been widely used to control stocks for many years within small and medium sized manufacturing companies. It is a pre-requisite to the effective extension to full material planning and control that a company has a reliable stock control system in place into which the new extensions of material planning can be interfaced.

On site installation and commissioning

For companies engaged in developing systems rather than products, such as process plant, instrumentation and control systems, communications and power distribution systems, a major source of lost time and cost is associated with problems which do not occur until the installation and commissioning phases. Likely problems include, on site shortages of parts or materials, inconsistencies in assembly drawings (misalignment, poor fit, clashes, missing connection), unclear drawings lacking adequate details, missing drawings and functional shortcomings in equipment. Clearly, the failure to maintain the integrity of the information available on site is a failure of overall project management. It comes about largely from errors or deficiencies in detailed design, planning, control and documentation.

Whilst no computer based system will overcome poor engineering work, the availability of more structured and controlled information emanating from the design and production functions can help to reduce the incidence of on site errors and consequent delays and increases in costs. For example, the ease with which scrap views on drawings can be generated without any additional draughting effort, the ability to detect clash conditions within a three dimensional model of pipework, ductwork or cabling, or the improved control of the drawing set which relate to the job by means of manipulating the documentation within a computer based document control system, are some examples of how the accuracy and consistency of information available to the on site installation and commissioning functions can be improved.

The installation and commissioning function can also benefit from the system having been rigorously evaluated at the functional and operational level during the design stages. This should result in less remedial work required on site. Indeed, many of the advantages of greater integrity in the information released to the site, by using the disciplines required by computer aids, are now being perceived by the on site installation and commissioning functions. The pay off using computer aids at the earlier stages of design can begin to be seen when the system is processed to final implementation on the customer site.

Inspection and testing

The inspection and testing functions are crucial elements in the full design and manufacturing cycle. The achievement of adequate quality standards ultimately rests upon the inspection and testing activities. Problems sometimes occur when inspection drawings or testing procedures do not accurately relate to the design and manufacturing information which have controlled the product or system up to the completion of the manufacturing process.

The direct extraction of inspection geometry from design drawings has been referred to earlier and is widely used within a computer aided engineering environment to ensure compatibility of test/inspection drawings. Within the design of mechanisms it is also possible to extract toleranced geometries of the components of the mechanism. These can be used in key positions within the range of movements of the mechanism. This information can be a powerful help in the inspection process.

On-line gauging during machining activities is capable of being handled in a similar manner to numerically controlled machining operations outlined earlier. Such gauging facilities can be added to machine tools and the necessary commands for gauging can be included in the numerical control part program.

Image processing in inspection

By combining television technology with computer based equipment the whole area of image processing has evolved over the last 15 years or so. The basis of this approach is to capture video frame images of an object and to convert this image into a statistised image in which each picture element (pixel) is characterised by interpreted data such as its co-ordinates, grey-level (light intensity), and possibly colour hue. By further processing this statistised image and comparing it with standard references it is possible to automate, to some extent, gross inspection functions. This approach is of particular benefit when inspecting items which are produced or processed in large volumes. Traditionally such items are inspected using either low skill inspection staff who visually inspect the passing items or by on-line simple inspection mechanisms. The former are often prone to errors due to boredom or fatigue and the latter to jamming.

Inspection techniques using image processing technology are likely to be further developed by including the emerging artificial intelligence (AI) software techniques in order to refine the decision making process based on the captured video image. Similar techniques are emerging for deploying image processing and AI within the handling systems associated with robot based manufacturing technology.

Printed circuit boards

A major area of computer based testing has developed in the electronics sector concerning printed circuit boards. In the first instance it is necessary to inspect to ensure that all the correct components have been inserted in the correct positions and with correct orientation, that all tracks within the printed circuit board design are present, and that all components are generally in working order and approximately of the correct value or type.

In order to go on to test the functionality of the circuit, complex digital electronic circuits require an extensive series of tests in order that all functional combinations and sequences of the circuit are checked. Efficient testing requires that the sequence of external input stimuli to the circuit and the observed output monitoring points should be optimised. This is mandatory in order that components and connections within the circuit are not covered more than the minimum number of times necessary to confirm the correct performance of the circuit.

For both these types of tests the data required to determine both the input supplies and signals and expected output signals are implicit in the circuit information required to manufacture the product. If the circuit is designed using computer aids, the circuit components, connectivity and printed circuit board (PCB) topology are held within the computer models of the circuit schematic and PCB layer topology models. The functionality of each of the components is also held within the computer in the information held within the computer's component library.

Thus, software has been used for many years in order to generate programs and test pattern sequences required to drive automatic test equipment (ATE) in the optimum sequence. Whilst such programs can be written manually, this is time consuming of skilled manpower and error prone. Test patterns that test pattern sequences are often difficult to optimise manually. The largely automatic synthesis and generation of test pattern sequences approach to ATE programming can help to justify the use of ATE. This can be true, even where the expected volume of circuits to be built and tested is lower than that which warrants the investment costs of manually developing the ATE program.

Maintenance

The optimum performance of any product, piece of equipment or system can only be maximised if adequate and timely maintenance procedures are undertaken. Maintenance requirements normally demand some combination of planned maintenance and breakdown maintenance. It is also a pre-requisite that adequate manpower, and capital equipment resources are made available.

Computer aids for maintenance have emerged in order to schedule the maintenance operations for plant, products in the field and the equipment and skills required to be used. This activity is analogous to the production planning and control operations highlighted earlier in this chapter. Here, even such basic techniques as the use of simple computer based spread-sheets can be a powerful aid in exploring 'what if' situations. The use of manpower and associated maintenance equipment, the maintenance downtime of plant or equipment, calibration histories, cost implications and spares holding strategies can all be explored by relatively simple general purpose computer tools.

More sophisticated computer aids aimed at the maintenance function have been developed and in use for over ten years - particularly in such areas as the petrochemical industries, chemical processing, transportation, food processing and nuclear industries. Here the computer aids have been used to control planned maintenance resource scheduling, record keeping, fault analysis, spares management and cost of ownership over lifetime analysis.

In the areas of unplanned and breakdown maintenance, computer aids, based on processing structured sets of documentation held within computer memory and conventional programming techniques, are used as aids in diagnosis and remedy of faults. As outlined earlier, the use of software techniques associated with automatic test equipment within the electronics industry has been of great assistance with handling the testing of electronic based systems of ever increasing complexity. This technology has had even more impact on the maintenance function. The logical approach which is implicit allows the computer aided system to identify the small subset of components or connections which are most likely to be causing the observed fault. This information can greatly speed up remedial action.

Similarly, customised computer aided systems have been designed to aid maintenance engineers. The earlier versions of these systems were based on some form of logical model of the equipment, together with a set of rules which relate to input stimuli or observations with output measurements and observations in order to identify likely causes of failure. In recent years such maintenance aids are now emerging in which AI software techniques are used. This type of maintenance aid can not only infer the likely cause of the problem but also has mechanisms for increasing the rule base on which diagnosis is founded by dint of 'its experience' and interaction with the expert users of the system.

As built documentation

In many situations where the equipment or plant is installed within a site, it is necessary to record the 'as built' situation for future use. It is often the case that at the end of building an extension to chemical process plant or conveyor system, the precise layout detail, as originally proposed, is not implemented on site. Local site conditions, which may have been unknown or not recorded at the start of the development cycle may cause modifications to be made at the time of installation.

The drawings and records of the plant often need to be updated to the actual on site conditions at the end of the project. This requirement may be a legal obligation or may simply represent an opportunity to update the accuracy of data concerning the plant. Clearly, if the drawings and schedules of the plant are maintained on a computer based system, the flexible and efficient editing facilities which are associated with such computer aids as design and draughting can assist in rapid updating of drawings to the as built status.

A technology known as photogrametory has been under development for over ten years in order to assist with semi-automating the process of capturing the 'as built' equipment or plant design. This approach has been pioneered within the petrochemical and chemical process plant sectors. It represents a further development of the use of image processing techniques.

In outline, statistised images of the built plant are captured from a number of different view points together with known datum points clearly identified. These images are processed in order to interpret the three dimensional topology of the plant represented by the multiple statistised images. The interpretation is in the form of a set of primitive volumetric models such as orthogonal boxes, cylinders, spheres, hemispheres, cones, frustrums and toroids. Each primitive volume is assigned size and orientation parameters. These primitive volumes are similar to those used in one of the forms of three dimensional volumetric modelling used in computer aided design systems.

This interpretation process within the photogrametory system results in the formation of a primitive volumetric model of the plant (as built) which can be manipulated and subsequently worked upon within a computer aided modelling system.

This process is only now emerging from the research field and it is unlikely the technology would have direct application within a small to medium sized company. However, small to medium sized companies may, within the foreseeable future, be faced with a three dimensional 'as built' computer model which has been generated using this new approach as a starting point for further design.

4

Overview of the system components within the computer aided technologies

Within this chapter the basic elements which are combined to provide computer aided systems for engineers will be outlined. It is not the purpose of this overview to give a rigorous analysis of computing technology since this is available elsewhere in specialist books on general computing. It will suffice to introduce the reader to the main computing components available. These are combined to produce effective computer aided systems for supporting diverse applications for engineers and technical staff. These systems are now being used in a wide range of company sizes and structures. Special emphasis will be given to those aspects of the technology which tend to be particularly important to the use of computing technology in support of engineering applications.

In Chapter 1 the development of the underlying technology components was outlined and these developments have, to a large extent, controlled how computer based aids for engineers have evolved. There are three fundamental topics to consider further at this stage. These are:

- Computer hardware;
- Computer software;
- Application data.

Computer hardware

Most readers will be aware that the term 'computer hardware' covers the computer processor and the various devices connected to it. These are often referred to as peripheral devices and are necessary to service the following functions:

- Storage of data;
- Input of data in various forms;
- Output of data in various forms;
- Transfer of information between processors.

In order to service the range of aids for engineers within any particular company, careful consideration has to be given to the hardware configuration of the computer system. Modern computer aids for engineers can be supported by three generic classes of computer hardware system:

- The centralised processor based system;
- Engineering workstations;
- Personal computers.

Centralised computer systems

The centralised processor(s) based hardware system is characterised by having a central processor complex which may consist of more than a single processor. This centralised hardware services each of the engineering users around the company. All of the application software and the company data resides on this central complex. The user accesses the computing resource typically through a so-called 'dumb' terminal device which may be a simple alpha-numeric terminal (with no support for handling graphical information), or more likely a graphics terminal which allows both alpha-numeric and graphical data to be displayed on the screen.

The centralised computer approach tends to find favour in the following situations:

— Where there is a large number of users largely accessing the same set of company information;

— Where the applications are particularly demanding of computing power (e.g. simulation, synthesis, analysis);

— Where users are located over a number of locations which require them to access the computer via communication circuits (e.g. computer bureaux, education).

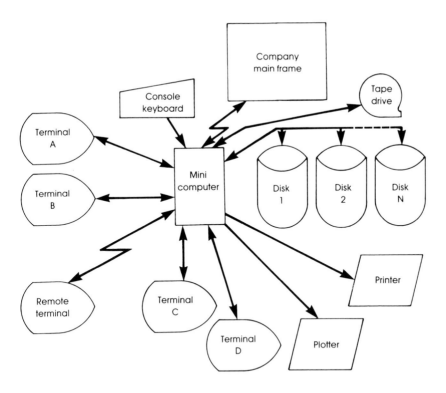

Centralised mini computer operating in contention.

With this centralised approach, the information storage devices, such as disk storage units, must be large enough to service all users of the applications on the system. Each user is allocated individual storage area on disk. Users are serviced with computing resources, typically by polling each user in turn and allocating central processing time as a function of the priority which that user enjoys. The overall management of the computing resources rapidly becomes a key issue in the administration of such centralised systems.

Disadvantages of the centralised approach

A problem with the centralised computer system is that in the event of any key element of the central equipment failing (e.g. power supply, processor(s), disk unit(s), communication equipment), service is lost to all users. This 'fail hard' characteristic can be overcome only at considerable expense by duplication of key elements within the centralised system and parallel operation of the duplicated equipment.

It was noted earlier that one principal limitation of this approach concerns the consistency of the response time of the system to the users. This is because each user is in contention for the centralised computing resource and dependent on the task which others are undertaking. Thus, if the range of tasks which are supported include a combination of highly interactive tasks (such as computer aided draughting) and highly computational demanding tasks (such as design analysis or building three dimensional computer models), user response times can vary alarmingly if precautions are not taken. The fall off of response time is typically rapid as the demand on the system increases, either by additional concurrent users, or by one user starting to use an application which imposes a high demand on the processor or the disk units.

The simplest way of avoiding the response time problem, with a centralised processor configuration, is to ensure that the computing power and speed of the central computing resource is high enough to service the worst case loading situation. This means maintaining acceptable response times for the highly interactive applications in this worst case load condition. This can lead to high initial cost for central processor configurations.

Another way of optimising the response time is to assign different computing priorities to different users, such that those which have a need for rapid response are serviced with computing resource more often or in large slices at each turn. Applications which are heavy users of resources often require less direct user interaction when they are running in their most 'hungry' phases (e.g. finite element analysis, printed circuit board automatic track routing). For such applications it is prudent to assign them to run in 'background' mode, so they are serviced by computing resource left over from the more highly interactive users. Similarly, plotting of output drawings can often be handled as a background task. Whilst those tasks which have been assigned to background operation will take longer to complete, they are perceived by the user as complex tasks and longer turn around times tend to be more acceptable.

Advantages of the centralised approach

The advantages of such a hardware approach include the centralising and control of engineering data throughout the company, centralised system administration of the system and the generally lower incremental cost of adding further users to the system - particularly if the user can be adequately supported by a simple 'dumb' terminal.

Engineering workstations

The engineering workstation has become the more normal work horse of computer aided engineering. The workstation consists of a computer processor(s) which is serviced by its own local disk and back-up tape storage devices, graphics display terminal, graphics tablet, communication facilities and hard copy output devices. Workstations are provided typically in families, with options on computing power, processor speed, random access memory size, spatial and colour resolution of the graphics display. They may optionally support data communications facilities, and often do.

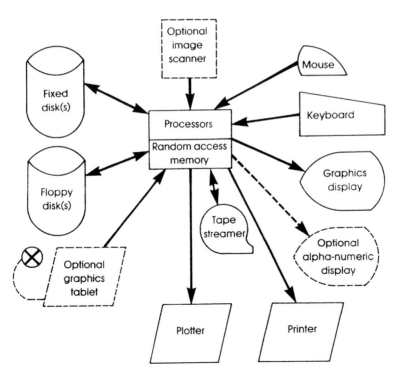

Engineering workstation hardware components.

A typical engineering workstation (courtesy of Computervision Corporation).

Advantages of workstations

The main advantage of the engineering workstation is that users have their own local computing resource. This can be tailored to meet the computing power required by the engineering applications which are predominantly used. Thus, if the workstation has been configured appropriately for the engineering tasks, the operational performance, in terms of response time to the user, numerical, spatial and colour resolution will be consistent. The application software and the data which is currently to be worked upon are held within the workstation's local disk memory. Thus the user is isolated from the effects of others. The user's needs can be serviced from this locally held information. The workstation can either be used as a stand alone facility or, alternatively, can be linked to other workstations, central computers or personal computers via data networking technology. The network can either be a local area network or a wide area network. The local area network services areas typically within a building or between closely adjacent buildings whilst the wide area network, as its name implies, services rather larger geographical areas. Networking is discussed later within this chapter.

Personal computers

Personal computers are now capable of supporting many (although not all) widely used computer aided applications for engineers. In order to do so they must be configured with adequate processing chips (typically at least 32 bit word length and running at 25 Mz. or faster), adequate random access memory (RAM) (typically greater than 2 megabytes) and sufficient disk storage capacity (say a minimum of 40 megabytes). Graphics terminals, tablets, printers and plotters can be serviced within the personal computer environment. The choice of such peripherals needs to be matched to both the main engineering applications and the cost constraints of the user company.

The upper end of the personal computer range when combined with high resolution graphics terminals and supported by adequate RAM and local disk capacity now overlaps the lower end of the engineering workstation range. The boundary becomes less clear with each advance in personal computer hardware.

Graphics display terminals

The development of graphics display terminals was briefly discussed in Chapter 1. Most computer aided systems in which graphical information is used now employ raster refresh terminals. The range of such terminals, from which the user must select equipment, is now considerable. The main features which vary are the physical screen size, the spatial resolution in the horizontal and vertical axes, the range of colours which are supported and the range of graphics control functions which are supported within the terminal hardware.

Physical screen size

Physical screen size is important and should be related to the application to which the hardware configuration is to be applied. For most computer aided applications which have a dominant graphical content, such as modelling and draughting, a minimum of about 20 inches diagonal should be used. However for applications where the graphical content is rather more by way of confirmatory information, such as graphical numerical part programming or process planning, the screen diameter can be less, down to 14 inches diagonal for example.

The spatial resolution

The spatial resolution determines the accuracy and clarity of the screen image. The lower or poorer the spatial resolution of the screen, the more jagged the shape of lines which do not fall on horizontal or vertical axes of the screen will appear. Curved lines appear as rather rough lines with 'staircases' if the spatial resolution is too low. For reasonable quality graphical presentation on a 20 inch diagonal screen, orientated in 'landscape' configuration, spatial resolution should not be less than about 1000 picture elements in the horizontal direction and about 750 in the vertical direction.

As the physical size of the screen increases so must the spatial resolution increase if the perceived quality of the graphical image on the screen is to be retained. The higher the spatial resolution of the graphics screen the greater the display buffer memory must be (invariably now held within the graphics terminal) in order to refresh the raster screen.

Range of colours

The ability of the graphics terminal to show colour is a further option which users must consider when choosing the hardware configuration for their system. The number of colours which can be used simultaneously on the screen depends upon the hardware specification of the terminal. At the bottom end of the colour terminal range eight colours, each of which can be flashed on and off, are supported. These are the normal range of colours used in teletext applications. For more advanced colour terminals colours can be simultaneously displayed on the screen and these can be mixed, both in terms of hue and intensity, using a colour palette facility. This facility enables the user to optimise the colour rendering on the graphical image. For most applications in which colour is a dominant requirement a maximum of about 4000 colours is adequate and is within range of graphics terminal technology.

For many applications, the use of colour is a subjective matter for the user. Some applications demand the use of colour graphics terminals. For example a computer aided system for designing printed circuit board tracking is not really practical with a monochrome screen. However an eight colour limit is perfectly adequate for the user to discriminate between graphical data of different types (e.g. tracking on different layers, power tracking, component pads). Similarly, for mechanical drawings, many users find advantage in being able to display different aspects of the drawings in different colours (e.g. construction lines, dimension lines, text and mechanical component boundary geometry). Much can be done to service such basic needs of colour with an eight colour capability.

There are a number of applications in which colour rendering is much more demanding. Such applications as textile or wall covering design, demand that the graphics terminal can resolve a large range of colours, both in terms of hue and intensity. Similarly, at the conceptual stages of design of many items such as clothing, consumer goods or professional equipment, if solid or surface models are to be displayed to their best advantage, the terminal should support a wide range of user mixable colours of different hues and intensity. Another range of computer aided applications where strong support for colour rendering is required is in cases where entire scenes are modelled (examples include process plant design, highway and building design).

The colour hue and intensity of each picture element has to be held in a further section of buffer storage memory. Colour graphics terminal technology has now been developed to a high degree. Coupled with well developed visualisation software, extremely realistic computer images are now routinely generated.

Graphics control functions

A further refinement of modern graphics terminals concerns the extent to which the terminal hardware can directly support a range of graphics manipulation commands from within the terminal hardware itself. Examples of the types of hardware commands which are now often supported within the terminal electronics include panning and zooming within the graphical image. Also, the ability to dynamically drag sections of the graphical image around the screen in real time, is important when exploring different design alternatives. The ability to view different aspects of parts of a drawing or image at different parts of the windowed screen at different scales can be a particularly powerful aid when exploring such aspects of mechanical design as the alignment and fit of components.

Tablets and digitisers

The electronic tablet is a device for transmitting a pair of X and Y co-ordinates by probing the flat surface of the tablet with some form of pointing device (typically a cross-hair puck, or stylus). The flat rectangular surface of the tablet can be annotated with a series of overlay templates. These are divided into different areas, each of which can be annotated either with a cryptic abbreviation for a command or by a simple graphical symbol or icon representing an application command. The overlay is similar to a restaurant menu and thus normally known as a tablet menu.

The method of operation of the tablet in a typical application consists of calling up a sequence of application commands shown on the overlay. This is performed by pointing the device on to the appropriate area of the tablet menu in an acceptable sequence to achieve the operational task required. This approach enables the tablet to be used for a number of engineering applications, by using a different overlay menu template for each main application. When an overlay template is changed, the tablet must firstly be initialised by registering the position of the application template sheet on the tablet (by pointing to two datum points on the overlay with the pointing device). This is done after selecting and identifying the application system which is required.

Each time the pointing device is activated a pair of X-Y co-ordinates is transmitted to the computer which is interpreted by the communication software (associated with the tablet) generating a string of command characters commensurate with the area of the template probed. These are in turn decoded to initiate the software functions achieving the required result for the particular application and the menu overlay in use.

Screen menus

An alternative to the tablet menu approach is to use screen pop-up or pull-down windows to hold similar cryptic commands or icons and to slave the screen cursor to the movement of a similar pointing device which is moved over a flat surface - typically a roller ball mouse, joystick or thumb wheel.

Plotters

The output from the computer system of graphical information such as drawings is achieved on a plotter. A wide range of plotters have been developed over the years. They range from simple A4 size upon which single A4 sheets can be plotted, to large and expensive plotters upon which full sized lofting templates for large aircraft parts can be output at actual size. The choice of plotter depends on the type of engineering application which is to be supported by the hardware system.

Three generic types of plotter can be considered. The first type is the most widely used, namely the wide family of pen and paper plotters. The second is the electrostatic type of printer/plotter and the third are specialist types used in plotting directly on to photographic film.

Pen plotters

Pen plotters can be of the flat-bed type in which single sheets (from A4 to A0 size) of drawing paper or film are loaded on to a flat surface and held there by a partial vacuum whilst the pen moves both in the X and Y direction carried on a movable gantry, along which the pen carrier travels. The pen plotter can also take the form of a drum type plotter (of various widths). In this type either single sheets/rolls of drawing film/paper rotate to and fro, held normally by friction rollers on the surface of a drum, whilst the pen is carried on a gantry along which it travels left and right in the direction perpendicular to the rotation of the drum.

More than one pen position is normally provided in the pen carrier. This enables pens of different thicknesses for various width lines or of different colour to be selected. The pen is made to rise on and off the paper as the paper moves underneath. The plotter driver software commands the pen, drum and pen holder movements, thus committing the master drawing to paper or film. This master can be reproduced by conventional methods such as dyeline printing or photocopying. The choice of drawing paper, film and pens is crucial. With irregular plotter use and poor maintenance, the pen can intermittently dry up causing parts of the hard copy drawing to be missed out in the plotting process.

The cost of the pen plotter is determined by operating parameters such as the spatial resolution (the smallest increment by which the pen or paper can move in the X or Y direction), maximum speed and acceleration of the movement of the plotter, maximum size of drawing film/paper which can be handled, number of pens which can be held and the robustness of the mechanisms which is determined by the expected duty cycle.

Electrostatic plotters

Electrostatic plotters are offered in varying widths and spatial resolutions.

They rely on a digital image of the drawing being generated and stored electronically prior to outputting the statistised image. The resolution of the digitised image is a constraint in the final quality of the image produced by the electrostatic plotter. The speed of output is much quicker than a pen and ink plotter and is independent of the amount of lines on the drawing once it has been converted to a statistised form.

This type of plotter can use conventional electrostatic printing principals similar to that used in reprographic copiers or ink jet printing technology in which inks (optionally of different colours) are first split into droplets and then steered by charged plates on to the paper in order to create the image.

Electrostatic plotters are however considerably more expensive than conventional pen plotters for any given specification of paper size and spatial resolution. They can now produce good quality copy and their use is indicated where large volumes of drawings are to be generated by the computer aided system. In this case the increased speed of output is needed to service the volume.

Photo plotters

Photo plotters are required for applications where the hard copy output of computer graphics is required to be on stable photographic film. The main need for this is found in the design of printed circuit board artwork in the electronics industry where the precision artwork is in fact a key manufacturing tool. Further applications for photo plotted output include the production of silk screen masks, instrument scales, labels and inspection templates. For many small and medium sized companies it is often difficult to justify the capital cost of an in-house photo plotter and the services of a photo plotting bureau are normally used.

Networked systems

For many companies, computer aided techniques are required to service a number of functions within the company, e.g. the design office, drawing office, production planning, production control. A number of workstations are required to service these various functions. Indeed, within a drawing office or other department there is often a need to provide a number of workstations, personal computers or other terminals to handle the workload. Information is often required to pass between workstations or other terminals. In systems which are not serviced by a centralised computer, (i.e. distributed processor based systems) it would be necessary to transfer data physically between the various workstations or personal computers if the processors were not able to communicate electronically with each other.

To overcome this operational difficulty a range of communication networks have evolved. Typically, an information communication system known as local area network (LAN) is used to allow communication between the various processors within a distributed system. Each workstation is connected to the network through an electronic gateway which handles the communication protocol to send and receive information over the network. Each access point to the network is assigned a unique node address for the network.

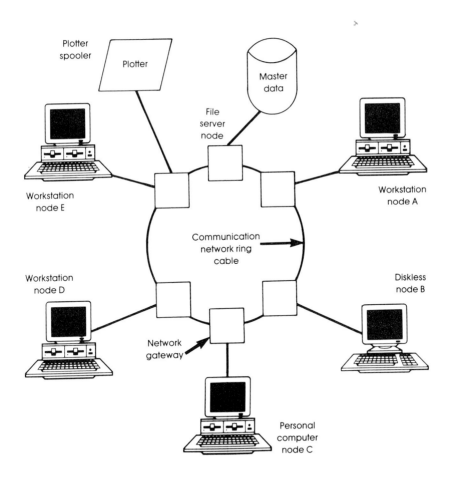

Networked architecture.

Information to be transmitted over the network is packaged up into blocks of digital information together with its destination address code and computed error detecting codes before being sent to the network. The receiving node on the network identifies the packets of information which are meant for it. After checking the error detection information within the package of information, it transfers the package of information into a buffer area within its memory.

Complete files of information can thus be transferred between nodes on the network in a series of information packages. The information so transferred can be optionally written out to a disk file at the node equipment for later local access and possible modification by operations carried out on the particular node.

Potential problems

With such a networked computing environment, there are potential problems associated with a growing number of files evolving at different nodes on the network which contain similar information. Clearly, a company using a networked hardware configuration must impose strict disciplines on the users of nodes. Included within these disciplines are the privileges which users have with regard to numerous files held at various locations within the network. Some files may be protected, such that they can only be read by certain nodes and not modified at that node. Other files may only be accessible on the network to users who have been issued with the appropriate password, protecting the file from access. Other files can be available to users who can both read and modify them.

In order to regularise the holding of important master file technical information held on the local area network, a preferred solution is to assign one or more nodes on the network as a master file server node. This processor node has direct access to a large volume of disk storage on which master files of data are held. Candidate master files include the following types of data which have been produced by a variety of computer aids used within the company:

- Master copies of computer aid application software;
- Master libraries of company component data;
- Authorised and released product or system information:
    - Performance specifications
    - Simulation and analysis results
    - Spatial models
    - Simulation models
    - Assembly drawings and diagrams

- — Detail drawings
- — Parts and material lists
- — Connectivity data files
- — Specification documents;
- Authorised and released production information:
    - — Manufacturing synthetics
    - — Process plans
    - — Part programs
    - — Production plans
    - — Factory paperwork;
- Authorised and released test information:
    - — Inspection drawings and templates
    - — Automatic test equipment programs
    - — Test specifications
    - — Test results
    - — Calibration results.

Individual technical staff within the company will generally take a copy of the data upon which they wish to work from the file server node, on to the disk storage device attached to their own node. Having worked upon the copy within the constraints of assigned access, the resulting data file is again held on their local disk file storage. At such time as appropriate data files are authorised, under appropriate password control, they can be transferred back to the master data held on the file server from which other network users can gain appropriate access to them.

The centralisation of master company data, together with released information relating to products and systems, goes some way to imposing discipline within the networked computer environment. Most users of the network will call on the file server to provide them with information which they wish to use with the local processing power provided within their node on the network. They can also transmit their own information held on the disk attached to their node to other users of the network directly instead of via the file server. This is often advantageous during the processes when the information being worked on at a node is still under development when preliminary information is required by another user of the network.

The rate of transmission of data around the network is of course a limiting factor on the time taken to transfer files between nodes on the network. Engineering data files are often much larger than most of those used within conventional commercial data processing. This is particularly true for drawing files. Even a drawing of modest complexity may consume of the order of half a megabyte. Larger, well populated drawings and complex three dimensional models may consume several megabytes of memory. Users of the local area network must contend with each other for data package time slots on the network, just as users of a centralised computer based system must contend for the processing power of the centralised computer. However, once the data has been transferred from another node or a file server node to the user's local disk storage, which may take from a few seconds to a number of minutes, the user has the full resource of the processor within his/her node to work upon the data. This helps to provide consistency of response time for the various functions he/she wishes to carry out on the data.

If the nodes of the network are separated by more than a kilometre or so, wide area networks may be invoked in order to network nodes together. Alternatively, remote access to a network node may be provided over a conventional modem link to the more remote terminals.

Software components

The operational functionality of the computer aid is, to a large extent, determined by the software within the system. The software content can be considered under the following three key aspects:

- Operating system software;
- Application software building tool software;
- Application software modules.

Operating system software

The operating system software is an integral part of the hardware supply. It is concerned with the routine management and housekeeping associated with various elements within the hardware system. System administrators, concerned with the computer system which supports the computer aids within a company, need to have a good general knowledge of the commands available within the operating system. They should undertake some formal training in the system which is installed on their hardware.

The operating system controls a number system parameters and general tasks of which the routine engineering user of the computer system generally remains blissfully unaware. These tasks include the following types of activity:

— Allocation of computer memory to the various tasks which are active at any one time;
— Organisation of file relationships and their subsequent handling;
— Prioritisation of tasks to be performed;
— Transfer of information between the computer processor memory and the various peripheral devices connected to the processor;
— Setting and handling protection facilities;
— Customisation of computer to national conventions;
— System debugging aids;
— System access and user usage monitoring.

If the computer system has been implemented as a networked configuration, network control software is required to be implemented. The networking system is an extension to the operating system and provides for the control and administration of the communication of information throughout the network. The networking system also needs to be understood by the computer system administrator within the company, to the extent that the network can be re-configured and its performance monitored and optimised.

To a large extent, the normal engineering user needs only a basic knowledge of the operation of the operating and networking systems e.g. logging on and off the system, setting up and structuring files, copying them from one location to another, taking security copies on to different media and communicating over the network if one is implemented.

Application software tools

The main aspects of the functionality of the computer aid is embodied within the application software. In developing application software a number of software tools have evolved and these tools are used within the range of application software modules which a vendor develops. Examples of software tools include:

- Database systems;
- Language processors;
- Graphics handling software routines;
- Data communication routines;
- Spatial modellers;
- Artificial intelligence routines.

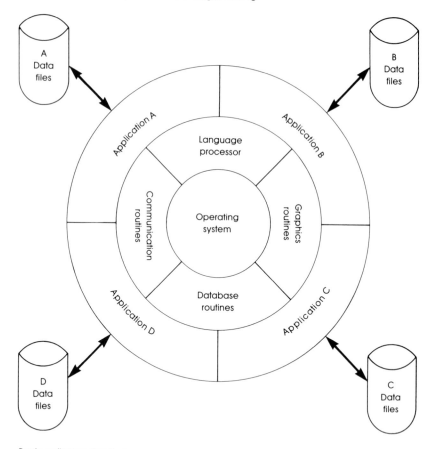

Basic software structure.

Database system tools

Such tools provide a rich development environment for the vendor's computer aids. They allow application software to be written using standardised methods of handling a number of aspects of the operation of the application module. For example, a relational database system can be used to format and model the relationships between different kinds of data which must be handled within a computer aided application. It also provides efficient and flexible methods for searching and manipulating the data under the control of the application software.

Language processor tools

Engineers wish to communicate with the computer using engineering terms and language with which they are familiar and which relate directly to the application task being performed. This language will need to have its own semantic and syntactic rules. For example if a piping engineer wishes to define the connectivity of the piping components within a piping branch run he wants to be able to use terms which he would normally use (e.g. CONNECT, ELBOW, ORIENTATION, LENGTH).

Whether he is using an interactive graphical method or a keyboard language to enter the details of the pipe run, the implicit nouns, verbs, adjectives need to be structured in such a manner that the rules for piping can be modelled within the structure of the application language which he prefers to use. This language needs firstly to be modelled in terms of its syntax and semantics within the computer, using some form of logical statement of the allowable structures and relationships of the components of the language. This engineering language must be converted into a sequence of instructions which can be followed by the computer, in terms of the calls to routines, procedures and library records to be used and executed by the computer.

In order to facilitate the rapid development and modification of such language structures, the concept of a language processor (sometimes known as command processor) tool has evolved. This allows the particular engineering language to be defined in terms of the rules allowed by the language processor. The engineering language can then be validated against these rules. Following validation, it can be converted by the language processor into the necessary computer commands. This process is invoked when the commands are applied within the engineering application by the engineering user.

Flexible and efficient language processors can significantly influence the ease of use and performance of the particular computer aid. They also allow the developer of the computer aid to optimise the structure, semantics and syntax of the language which will be used either explicitly or implicitly within a computer aid. These facilities enable the resulting engineering language and *modus operandi* for the computer aid to be tuned to meet the operational needs of the practical engineering user.

Graphics handling routine tools

Much of the communication between engineers is in a graphical form. Engineers and technical staff generally feel comfortable with graphical information. Many computer aids allow the user to interact with the computer in a graphical manner. Computer graphics tools allow the developers of computer aids to incorporate such graphical methods into the operation of their application software. The use of the following graphical facilities was identified earlier in this chapter:

— On-screen structured pop-up and pull-down application menus;
— Overlaid screen windows;
— Menu icons;
— Graphic tablet menus;
— Scanned graphics picture capture and manipulation.

Each requires flexible software tools to allow application software to take advantage of the hardware technology. Selection of commands using on-screen command menus, which can be called up in a structured manner, and which relate to the operation of the application is now commonplace. This removes the need for the engineering user to key in the commands directly on the keyboard. The structured menus also indicate the command options which are available at any stage within the operation of the computer aid. As each command level within the system is invoked, the sub-option commands can be displayed on pop-up menus overlaid on the screen. Application software tools are needed to set up the menu structures and to interpret them.

As an alternative to on-screen menus, some users find it preferable to select their commands using a graphics tablet, the operation of which was described earlier. Software tools are needed to register the datum points of the tablet overlay template and to interpret the X and Y co-ordinates which are transmitted when a command is selected using the tablet.

Some computer aided applications benefit from the use of graphical icons for inclusion within menus. An icon is a picture which relates to the command or user chosen option in a direct graphical manner. For example, the 'draw' command within a draughting system could be represented by a picture (icon) of a pencil. Icons can also be useful for identifying such things as components within a computer library. For example, an icon consisting of the standard graphical symbol for an electrical resistor can be shown on a library menu either on the graphics screen or on a menu template which is mounted on an associated graphics tablet. Such icons must be chosen carefully if they are to be helpful to the engineering user. Some users find them more confusing if they are not familiar with the computer application. Again software tools are required to allocate icons to commands and to access associated library items.

In addition to the graphics tools associated with command sequences, graphics tools have evolved for assisting software developers in creating graphical elements within their application software. Such routines take care of drawing lines, arcs, circles, polygons and filled in areas. They consist of a number of programming sub-routines and generally handle the transfer of graphical information to and from a wide range of graphics type devices such as graphics screens and plotters.

Scanned graphics

In addition to graphics which are computed using graphics commands, another form of graphical information can be of use within a computer aid. In this form, the graphical image is represented by a series of picture elements which are set to a given level of intensity and optional colour attributes. Such images can be scanned into the computer memory by a scanner device and held in the form similar to that used for newsprint pictures. The scanned images can be stored on disk and recalled for use later within documents produced and edited on such computer aids as desktop publishing systems. Software tools have evolved to manipulate such scanned graphical images. Scanned images cannot be interpreted in the same way as engineering drawings which have been generated by a computer aided draughting system. They are held purely as picture elements and no structure in terms of lines and text is modelled within the computer.

Data transfer tools

The use of network communication systems for sharing data within a number of engineering users was discussed earlier. There are however aspects of data communication between users of computer aids which have come about from the historical development of the technology. The different manner in which engineering information is modelled by different vendor's computer aids gives rise to the problem of communicating engineering data between two computer aided systems from different vendors. Software tools have evolved to try to overcome these difficulties.

A classic case of this problem relates to the modelling conventions used to model graphics data which has been computed by the computer aided system. Two approaches have been followed. The first is to formulate a common internationally (if not always intercontinentally) agreed neutral standard containing a number of agreed options for modelling of graphical elements and associated data. Each vendor's system then offers software tools to output the data into a neutral standard format and to input from it. This approach has met with some success but there can still be problems in editing the data further when the data has been imported using the software tools to read from the neutral format. It is often necessary to know which of the standard modelling options have been invoked by the originating computer aided system in order to maximise success.

The second has been to build specific interface tools between the more widely used vendors systems. This approach can be more successful but at the cost of limiting the use of each system to the functions and constructions which are supported by both the two dissimilar systems and the constraints of the interface software tool. Both of these methods are discussed further in Chapter 13.

Modelling tools

In order to model the physical aspects of a product or component, a number of important modelling tools have been evolved. An engineering artefact or component can be modelled in a number of ways and software tools are now available to enable the engineer to model these physical attributes. The main forms of computer dimensional modelling can be summarised as:

— Modelling of multi-sheet two dimensional plan and elevation drawings;
— Wire frame modelling of three dimensional objects;
— Full solid modelling of objects;
— Modelling of complex doubly curved surfaces.

These computer based modelling techniques are discussed later in Chapter 6. They are combined in a wide variety of proprietary application software packages which are targeted at the design and draughting office within a company. The use and application of these different modelling techniques need to be fully understood within the company before the company can make a sensible decision on which candidate commercial software to acquire. The combination of dimensional modelling tools which are required is highly application dependent. It is most important to consider how the use of computer aided techniques is to develop within the company in order to avoid closing off technology upgrade paths for the future.

In order to model systems rather than products, other computer software tools have been developed which model symbolic schematic diagrams. Such diagrams consist of symbols which represent components, subsystems and connection lines which indicate how the system is connected together. Typical examples of these types of diagrams include system block diagrams, electrical, hydraulic or pneumatic circuit diagrams and logic diagrams. The modelling tools which model these sets of schematic information are capable of modelling multi-sheet diagrams at different levels of physical implementation of the system (e.g. in the case of logic diagrams - at the gate level or at the level of the physical component package containing multiple logic elements or even higher levels). Such application software tools are in turn supported by appropriate symbolic component libraries. Additionally, they enable the user to extract connectivity information and component parts list. They also afford interfaces to other software tools. These include analysis and simulation aids or downstream computer aided tools for tasks such as topological design (e.g. printed circuit board or chip topology design tools). Interfaces are also provided to system testing aid software tools to support automatic testing in various ways for the system.

Knowledge modelling tools

The ability of computer aids to acquire, recall and apply knowledge in order to advise engineers is now possible. The use of knowledge based system tools offers an extension to the scope of traditional computer aids for engineers into the area of promoting best practice in engineering. The range of applications of these knowledge based systems using AI techniques is currently relatively limited but is set to expand throughout the 1990s.

These knowledge based system tools provide for the storage of application specific rules together with the underlying explanation of each rule and its dependencies. Often the technique is then to review application data (e.g. design, manufacturing or test data) which has been built up by the users of a computer aid, against the structured set of rules which relate to the aspect of the design to be reviewed. Such analysis uses the rule set together with the application data to infer when and where advice or warnings should be given to the user of the system. Such advice stems from the breaching of rules which are held within the knowledge based system in relation to the application data being reviewed. Rules can be structured and assigned priorities and the user can ask the knowledge based system for an explanation as to why the advice or warning is being given. In which case the system back-tracks through its rule structures and dependencies to provide the user with its 'reasoning' and the objective goals for which the system is striving.

The users of such system tools can also interactively request advice on a topic or option during work with the system. They can ask for a review of their data so far constructed within the computer model. They can override the advice or warning from the system but they do so knowingly and this will be auditable on the system. A reason for overriding such advice may be that the knowledge based system does not have a rule set which relates to a particular set of circumstances. For this reason such system tools must provide for an authorised user to input new knowledge structures in the form of further rule sets which are linked into the existing set of rule knowledge.

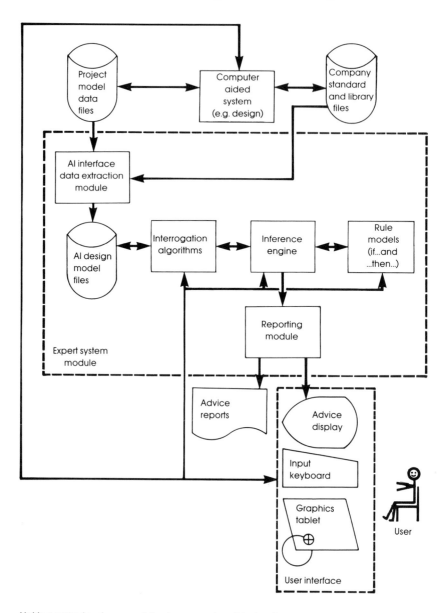

Linking expert system modules to computer aided system.

Thus, a credible knowledge based system must provide the following basic facilities:

— A data storage model for modelling the application data in a form which can be used in association with the rule set it is testing against;
— A model for storing the knowledge about the application in the form of rules;
— An associated model of the reasons and explanations which underlie the rules;
— A so-called 'inference engine' which infers from the application data which rules have been breached and provides the consequential warning or advice action to the user;
— An interactive mechanism for acquiring new rules from authorised users (known as the domain expert) and incorporating these into the existing rule set;
— An easy to use man-machine interface to facilitate the easy use of the knowledge based system.

These new tools are by their nature application specific. They are potentially beneficial where there is a code of practice relating to the activity to which they relate, or where technical information needs to be validated and optimised with respect to a particular company goal. Examples of engineering related knowledge based system tools which are in various stages of development include:

— An aid for computer system configuration;
— Cost estimating in a 'make to order' environment;
— An aid for feature recognition;
— Design for testability;
— Design for manufacturability.

Application software modules

By combining the various software tools which have been outlined above and application specific standard data libraries, with an appropriate hardware configuration, a set of computer aids can be configured to best meet the needs of the markets for different engineering applications. It is from this set of computer aids that a company has to choose and match its perceived needs for computer aids. However, to the inexperienced user of computer aids, it is not possible to assess the functional capability of a combination of tools purely from looking at a set of drawings or screen displays. Many drawings and

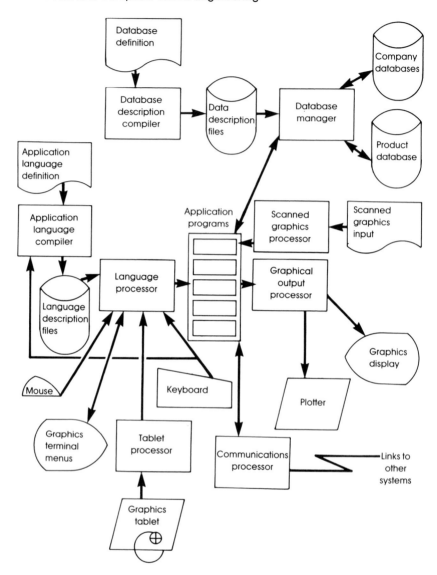

Use of software tools within application software.

graphical presentations can be produced by different combinations of software tools. What is important is not the picture on the screen or plotted drawing, but what was involved in producing them and what lies behind the data held within the computer from which they were produced.

5

Justification for introducing computer aided engineering technology

Within the previous chapters the scope and content of the computer based technology aimed at assisting engineering and technical staff have been discussed in outline. The decision to embark on using computer aids in technical areas or to extend their use within a company is one which can have a major impact on the company. It must eventually be made on a commercial judgement as to the financial and/or technical advantages which will accrue. Within this chapter the justification criteria will be considered in order to provide some guidance to company management. Managers must consider how to justify the capital investment involved within their own companies.

The potential benefits across the range of computer aided engineering techniques must be assessed on a case by case basis, especially within small to medium sized companies. Computer aids are not appropriate in all cases. The availability of a skilled member of staff, a pencil, a piece of paper, a set of drawing implements, a set of standards manuals and a pocket calculator is still a powerful combination within a company office. What is more, the capital employed to support such a conventional member of the engineering staff is quite minimal.

How then does a small or medium sized company set about trying to justify investing additional capital in order to support its technical staff? If it is proposed to spend more capital on a new machine tool or a piece of test equipment, the financial management of a company can relatively easily relate such expenditure directly to the business of producing goods for sale. These goods leave the company's premises resulting in cash returning to the company from its customers. If such capital expenditure results in a reduction in the direct cost of production and testing of the goods, by way of reducing labour costs or reducing scrap, the assessment can be relatively straight forward. The senior non-technical management, who often control the capital expenditure of the company, can simply consider the savings and pay back period. The factors can be set against the other opportunities for using the capital and the justification relatively easily assessed.

The prudence of investing substantial sums of capital in technical activities of the company which are further upstream from the manufacturing and testing functions can be more difficult. To the non-technical management these upstream functions within the company are less directly visible. The output from them is largely information in the form of drawings, schedules and other documents which are not easily understood or capable of being monitored or assessed by non-technical senior management.

Pressure for investment in the technology can come from the bottom up from technical enthusiasts for computer based methods. Alternatively, it can come from the top down from managers who have seen similar technology being used in either the company's competitors or within companies within which they have previously worked. Management perceptions may be formed even from the result of a visit to an exhibition, conference or a casual discussion on the golf course.

Assessing sources of benefits

To set about justifying any investment in technical computer aids, it is necessary to recognise that there are longer term advantages if the company plans towards taking a corporate view on the use of computer based systems for its technical staff. To leave the investment pressure to come from individuals within the technical functions of the company can lead to ill-planned investments, lack of commitment to make the technology work and to technology blind allies. These problems may not be seen until some considerable time into the future, by which time the investment could be wasted.

Initial assessment

As a starting point for exploring justification for a short to medium term investment, it is necessary to consider each of the technical areas within the company. Studies are needed to identify which, if any, are experiencing problems that could potentially be eased by investing in computer aided techniques. The following functions should be considered:

— Technical sales and tendering office;
— Design and development office;
— Design drawing office;
— Production drawing office;
— Process planning office;
— Production planning office;
— Production control office;
— Test department;
— Maintenance department;
— Technical publications office.

The flow of information between these various company functions should be assessed. The activity and volumes of documentation of each type produced should be estimated within each function. During this assessment, bottlenecks should be identified and sources of error in technical data flowing within the company sought. The quality of information, as perceived by the offices which receive it from other functions within the company, should be noted and particular problem areas identified. A common problem with such assessments of the quality of information is those who produce the information typically consider it to be of a high quality whilst those who receive it, as the starting point for their tasks, often perceive it as less than ideal, error prone and partially incomplete. For example, any frank discussion with the design office, concerning the information provided by sales and marketing or with the production department, concerning the information provided by the design office or the design drawing office, will often tend to demonstrate the problems of assessing information quality.

Such assessments depend on holding a wide ranging series of interviews with management and operational staff within the various technical functions. It is important that whoever conducts the series of interviews can gain the confidence of the staff from which they are seeking information. They must be above suspicion of being partisan. It is important that an honest assessment is documented without fear or favour to the individuals providing the information.

Some companies find that the use of external independent consultants can provide such a service. Alternatively, someone from another branch of the company may provide sufficient detachment from the day to day problems of the site and the foibles of the staff concerned.

Prioritising problems

From this initial assessment, problems identified can be prioritised within each of the functions. Those which may potentially benefit from the use of more (or less) computer aids may then be identified. This process enables candidate generic types of computer aided systems to be focused upon for further study and assessment. It also clarifies the company functions which would be impacted on by the investment in computer aided techniques.

Specific company functions suited to computer aids

The investigation should be concentrated next on the company functions for which computer aids may be of technical and commercial benefit. Their current methods should be examined further and the characteristics of potential computer aids needed to offer benefit to these functions documented. The impact of a particular computer aid in the simplistic terms of potential saving of man hours within a particular department can then start to be assessed. This can only be done by someone who has experience of the technology as it is used within a similar operational environment. For example, as will be seen later, the potential saving in draughtsman hours in a drawing office is highly dependent on the volumes and mixture of types of drawings which are produced and modified within the office. It is also highly dependent on the pattern of work and non-drawing tasks undertaken by staff within a drawing office. It is simplistic to assume that a typical senior or intermediate level draughtsman spends a high proportion of time actually drawing lines on paper. If such time is analysed it will be found that typically more than half the draughtsman's time is spent on the following non-drawing tasks:

— Searching for information;
— Liaison with design, production or inspection and test staff;
— Attending meetings within the company or with clients;
— Doing calculations;
— Checking other people's work;
— General administration.

It is well known that computer aids for simple draughting are cost effective and their impact is to improve the productivity of the staff using them! How do we know? The answer is simple, the salesmen tell us so! The salesman will often quote simple productivity gain ratios for different types of drawing. As with other applications of computer aids, using simple productivity ratios to assess financial viability of an investment is fraught with dangers.

Justification must be treated on a 'horses for courses' basis. So much depends on the application, the mix of skills and the functionality of the computer aid. Showing a simple clear case based on the comparison of direct costs is likely to be unsound. Investing in computer aids for the technical functions of the company is not like buying a new machine tool or piece of process plant.

A more rigorous direct comparison based on direct labour cost savings, assuming typical write down periods for the computer based systems and including the amortised cost of the equipment will produce, at best, a break even situation for many application areas. Others may show a direct cost balance in favour of manual methods. However, such a simplistic approach is often less than half the truth. Other aspects implicit in moving to computer aids may generate potential savings. These can have a greater impact than any direct labour cost savings. Of course they may not. It is still frequently the case that the mix or volume of work tasks within a small to medium sized company does not justify an investment in a number of aspects of the technology. It is unlikely however that none of the applications would be of commercial advantage to the company.

Within the rest of this chapter a number of less directly quantifiable cost benefits will be discussed. These must be included in the argument when seeking a justification for investment in computer aids for the technical functions of a company. The extent to which they apply for any company will be a matter for that company to determine. Each should be considered and a subjective judgement made - possibly with a financial benefit assigned to the introduction or extension of the technology.

Shortening lead time

The pressure to shorten development lead times was identified earlier. With a mature set of computer aids to support a subset or all of the functions within the design to manufacturing cycle there is a potential for shortening the lead time for a new product. This is particularly true where a high degree of standardisation can be achieved in the sub-assemblies or modules from which the new product or product variant is to be made. A 'pick and mix' environment is ideal for gaining the best advantage from a computer aided approach to design. However this is only the case when the system has matured within the company and the standard sub-units have been set up as company-specific data within the system. Clearly, more rigorous analysis at the early stages of design, either from an aesthetic or operational view point, helps to reduce lead times for achieving a successful product.

Much of the time lost in finalising a product is concerned with completing the frequent changes to documentation as the project evolves through its various stages of design, development and production prototyping. Computer based methods are particularly good at handling changes and retaining control over the other information which is affected by the change.

The potential for shortening product lead times may be the dominant justification for investing in computer aided techniques. To be late to the market with a new product may seriously risk losing significant market share with large consequences for the whole business. This is particularly true as seen in markets where fashion or seasonality plays a major part. It is also crucial in sectors where product technology is progressing rapidly.

Improving consistency and quality in technical documentation

A significant cause of rework and loss of time concerns lack of consistency and poor quality in technical documentation. Badly drawn drawings or unclear text can give rise to poor communications between technical departments. This can arise from using less than optimum staff when under pressure or lack of proper control over technical documentation. Whilst a computer aided approach will not make a good engineer or technician out of a poor one, the quality of lines and text which appear on documents via the computer system will be consistent and more readable. The integrity of the relationships between the various documents produced by the computer aided system also stands a better chance of being more accurately maintained.

Optimising the appearance and operational performance of a new product or product variant, to meet the needs of its market, within decreasing development timescales, is a key consideration in the success of both the product and company which makes and sells it. Engineers frequently complain that they could produce better products and systems if only they had more time to explore design alternatives. Ideally, exploration of options should continue to the point where a well considered decision can be made on the option to be chosen for full development through to manufacture. Again, one potential justification of computer based techniques is the potential for rapidly changing design styles, colours, textures, details, mechanisms and control facilities, exploring these alternatives to a point where a good decision can be made on what to proceed with. Of course, the notion of exploring more alternatives before committing the design must be reconciled with the aspiration of reducing overall development timescales.

If as a result of exploring more alternatives, a better series of products result, the potential growth of company sales is enhanced. This can lead to greater potential for profitable growth throughout the company. Overall, company growth is a major objective and its continued achievement can have a large impact on the case for justifying aid which facilitates it.

Companies who are engaged in designing and producing systems rather than products, often have to customise such systems to meet the client need. Computer aids have proved successful in enabling companies to improve the quality of their tender documentation. This can be achieved by giving better detail and visualisation of the proposed system to be supplied together with more accurate tender cost information. If better tender documentation can be seen to improve the sales 'hit rate' on tenders, this could be the most dominant of all elements in the cost justification for improved computer aided techniques. It provides a further engine for company growth, possibly far greater than the direct cost advantage in saving man hours in the company's technical functions. This would be particularly true for companies with a sales order profile characterised by a small number of high value systems or product supply contracts. In such companies, the potential of increasing success rate on sales tendering is a key issue and may well be the principal justification for using better computer aids within the technical sales function.

As was indicated earlier, skilled staff in a number of key potential growth areas continue to be in short supply. This, despite periods of recession (which generally only serve to shake out the 'old wood'). It is important for many companies to retain the skills which they have and provide efficient aids to maximise their output. There is an element within the justification for investment in computer aids which depends on being seen to support the company's key technical staff with state of the art tools to do their job efficiently. To fail to do so can lead to staff instability and disruption to the company in having to replace its scarce skilled staff. Some key staff of course will tend to run a mile at the thought of changing methods after many years. These tend to be in the minority. They can be very valuable acting as devil's advocates when evolving the justification for investment in such equipment.

Documentation revisions

A search of the typical technical documentation within a company often reveals that many documents go through five, six, seven or more releases within the period of optimising a product or system for the market. Often, these changes in documentation result from information which comes to light as a project evolves through the development cycle and through prototyping.

The consequent rework and changes to documentation cause extra cost and time delays in stabilising the product or system ready for the market. Many small and medium sized companies, if pressed, find it very difficult to state what the costs of design and development really are for any given product. Tracking of costs against a particular product through all of its design and production iterations is often inadequate. The arrival of computer aids can assist, not only in effecting the necessary changes, but also in providing audit trails for tracing the costs of the tasks involved.

Providing early exploration

A key target benefit of computer aided technology is the potential to undertake rigorous exploration of a wide range of technical issues at an earlier phase of the design and development cycle. This can take a number of forms, from thorough simulation and analysis, greater detailed exploration at the design and draughting stage, through to numerical methods in support of manufacture and rigorous testing and inspection techniques. For example, the ability to explore on the screen alignment and fit problems by extensive use of scrap views within the time available, prior to prototype manufacture makes for better quality of design, increasing the possibility of being correct first time. In complex system engineering, more comprehensive and detailed simulation of the system, together with the potential for automatic testing of the system (under each combination of input stimuli), both increase the probability of being right first time or certainly with fewer revisions of the design.

Within a number of industries, notably in the aerospace, automotive and process plant sectors, this potential for being right first time has helped to reduce the number of prototype stages which need to be undertaken prior to full committed production. The same potential can exist for small and medium sized companies provided that staff make constructive use of the technology. Being right first time is a key objective within any task. If it can be achieved, the cost advantages to the company can be large indeed. Again, such cost advantage can often be of the same order as any estimate of direct labour cost reduction from using computer aids.

Monitoring and control

Most companies have evolved some form of monitoring and control systems to record, co-ordinate, track and analyse the commercial data flowing within the company. Such basic controls are enforced on the company in order to meet statutory requirements for reporting commercial activity - accounts have to be prepared and bank managers have to be satisfied! However, for many small and medium sized companies similar controls do not emerge as quickly as the company develops. Much of the technical information is held in a less than structured way and much of the communication between the technical functions of the company is at best done through *ad hoc* discussions, meetings and memos - often by scribbled notes and brief snatches of conversation over a coffee break or along the corridor.

There comes a time when a failure in technical documentation and communication (either due to inadequate documentation or failure to ask the appropriate question of a technical colleague) causes a serious problem within the company. This can often lead to substantial financial consequences. This typically occurs when the company is trying to grow and such systems of communication and control in place within the company are no longer adequate to meet the volume and complexity of the flowing technical data.

Structuring and standardisation

Computer aids can be used to structure technical information, track the consequences of changes and modification and the impact of external inputs to the engineering activities of the company. Not only can technical information be recalled more easily and relationships and dependencies of information be established, where appropriate, subsets of information can be easily re-used from one job to another, reducing the tendency to re-invent the wheel.

The better control of technical information is not automatically imposed by the technology but if technical staff are motivated to improve this aspect of their work there are facilities within the technology to make the structuring, recall and re-use of existing company data easier. As with other tools, it is important that appropriate staff are trained to make use of these additional facilities.

A key way of controlling costs within the technical functions of a company is the use of more standard and rationalised methods of designing or manufacturing products and systems. Many companies seek to rationalise their product ranges and to try to evolve families of sub-assemblies, subsystems and components from which the range of the products can be configured and manufactured. If such a strategy can be achieved, the technical risks of new developments or customisations can be minimised. The stock holding requirements can be significantly reduced, hence reducing the working capital for the company.

Computer aided techniques can help to support such a move towards greater company standardisation and rationalisation. The use of standard components, families of parts and sub-assemblies can be made more attractive to the technical staff concerned by improving their accessibility within the associated computer based libraries. It becomes easier to use a company standard than to embark on an entirely new and non-standard component or method. Also, where components fall into families of parts, in that they are of a similar basic geometry but of different combinations of dimensions, these can be set up in a parameterised form on the computer system's component library. When they are invoked by the user, the required dimensions are assigned and the computer system will create the component modelled to the correct dimensions required for the application.

Numerical modelling

In some market sectors, the use of computer based numerical modelling techniques becomes obligatory, simply because manual methods can no longer cope with the physical resolutions and accuracies needed, or the depth of analysis required. The classic cases include large scale integrated circuit design and manufacture within the electronics industry. This demands the

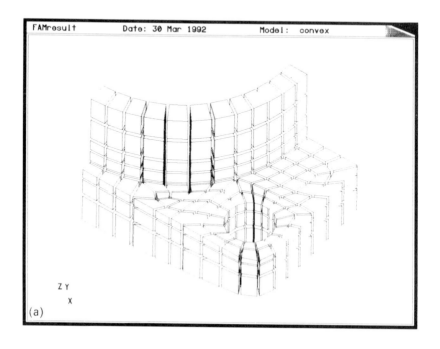

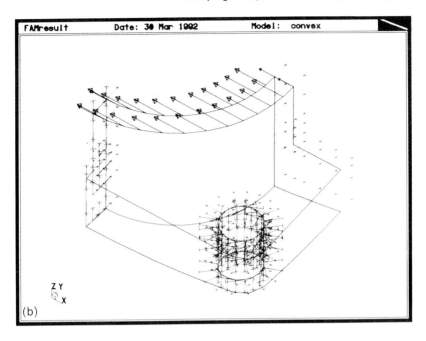

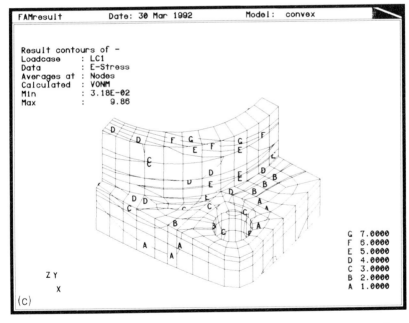

Finite element analysis of a component detail: a) 3D finite element model; b) The loaded and constrained component detail geometry; c) Finite element stress analysis results (courtesy of FEGS Ltd).

use of simulation analysis of complex circuits and high accuracy topological design. Finite element stress analysis used in aspects of structural design is a further example. Within these and similar cases, a key aspect of being in the business, is the level of investment which is mandatory in computer related technologies. With the trend of many western nations to move upmarket in their commercial activities, to provide the commercial opportunity for greater added value, there is the pre-requisite to strive for greater accuracy and quality in the products and system needed to service these market sectors. In such cases, the cost of being in many of these businesses is an obligatory increase in the investment of more advanced technologies.

Conclusion

It may seem to be special pleading to attempt to justify an investment in computer aids for the technical functions of a company, based on anything other than potential for direct labour cost savings. However, whenever an investment is being considered which is some way from the direct making and testing of products, wider issues have to be considered. The introduction of computer aids must be accompanied by a total review of the way in which the technical functions of a company operate, how they communicate with each other and how a more structured approach to handling technical information can impact on the downstream activities. Downstream that is from where the technology has been introduced.

Investments in such technology can now cover a wide range of costs; from a few hundred pounds to many tens of thousands of pounds for a small to medium sized company. Very low cost systems often serve only as 'taster' systems which hint at the potential of more comprehensive systems. They can also be useful in themselves if only to unlock the suppressed demand for more capable systems.

To allow piecemeal investment in technology packages within different departments can result in future incompatibilities and blind allies. This is not always the case, but if there is potential within a company for developing a medium to longer term strategy for deploying such aids throughout the company - in design, the drawing office, in production and test - it is important to evolve and document a firm management strategy. It is often the case that the value of company data committed to such computer based-systems, after a year or two following the installation of such systems, out-weighs the initial capital cost of the equipment and software. Ill considered use of inappropriate technology can place these cost expenditures at consider-able risk.

6

Using computer aids in conceptual design

The conceptual design of a product or system is normally the first stage of the design process and, as such, must be explored first before producing the documentation needed to build a pre-production prototype. These first activities in the design process are normally concerned with basic interpretation of market needs. The marketing function, where it formally exists within a company, is often instrumental in setting out an outline brief for the guidance of the design function. Initially, the design staff need to explore alternatives to meet this expression of market demands. Early consideration is made by the design staff of the following outline requirements of the new product or system:

— The external appearance;
— The main sub-assemblies and subsystems;
— The functional performance;
— The proposed methods of manufacture and assembly;
— The methods for testing and inspection;
— The maintenance needs.

Exploring external appearance

Traditionally, in order to design the external appearance of a new product, design sketches are made which are used to explore alternative design schemes. In the design of products such sketches are used to indicate main outline features of the product and identify how the various main parts relate to each other. Where the product is to be developed speculatively the design sketches normally stem from an outline specification provided by the marketing function of the company. Alternatively, for both products and systems, conceptual sketches may be prepared in response to an enquiry from a particular potential client.

As was indicated in Chapter 3, the design concept for the proposed system would normally take the form of an outline block diagram. This is used to identify the main subsystems, their operational functions, how they connect to each other and the principal inputs and outputs in functional terms. Also, within the system concept the various enclosures and physical relationships of the subsystem units would be sketched out. Outline details such as user controls, displays and alarms would also be sketched.

The main purpose of the conceptual design documentation is to communicate design options to the various interested parties both within the company and to third parties such as clients, sub-contractors or regulatory bodies. Where three dimensional models are made, these can take a number of forms. They can be made from solid materials such as wood, plaster, or wax or assembled from flat sheet components such as cardboard. Card models are assembled by bonding folded card components to represent the concept of the product.

Concept design sketches of a product are often coloured, using colour wash. Some attempt is also made to indicate the surface texture properties of

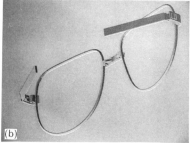

a) Computer generated solid model of assembled components; b) Exploring the relative positions of moving components by manipulating the solid model of the assembly (courtesy of Robary Ltd).

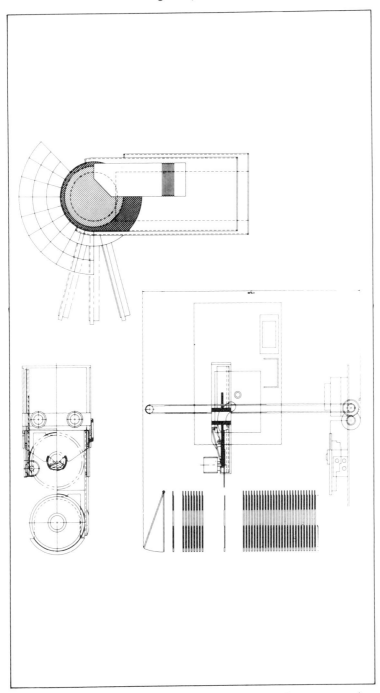

Exploring a compact disc loading mechanism design by moving component assemblies in different positions (courtesy of Warwick Evans Design).

the material. The sketches will typically show different views of the product and include three dimensional representations in order to facilitate good communication with non-technical staff. Many, if not most of these key people cannot read engineering drawings sufficiently well to gain a good appreciation of the appearance of the proposed product. In some cases it is necessary to indicate some of the proposed details of construction, particularly where some components or sub-assemblies interface with each other. Similarly some details are needed for such aspects as moving mechanisms. These details are difficult or expensive to incorporate into physical solid models in any realistic way - often partly because of the scale of the model.

Computer aided techniques are now sufficiently well developed to offer a powerful set of aids during the conceptual design phase within the design to manufacturing cycle. In the context of product design the emphasis has been on providing the engineer or industrial designer with the ability to explore their conceptual ideas at various degrees of realism on the screen. They can also pass outline geometry of their design to computer assisted model cutting technology in order to produce accurate three dimensional physical models cut out of modelling foam, wood or wax. This chapter explores some of the computer aided techniques available to improve the effectiveness of the engineering function at this conceptual design stage.

Even in a company which has installed computer aids within its technical functions, it is still likely that an initial design concept will be sketched on the back of the ubiquitous envelope, or beer mat when the enthusiastic engineer or industrial designer is struck by a spark of inspiration! After all, the best ideas often come when people are some distance from computer terminals. However, it is necessary to work up the idea into a set of documentation and other aids which will be adequate to engender confidence in the conceptual design within the company and beyond.

The computer aids available for exploring and evolving the physical outline of a product are based on developing and evolving some form of model of the product within the computer. A number of computer modelling techniques are available. The use of a particular type of modelling technique depends on the nature of the application, the sophistication of the computer aids implemented within the company, and the familiarity of the engineering staff with the technology.

Design sketching and modelling

Engineers are very familiar with representing mechanical objects using a series of two dimensional drawings of the conventional orthogonal views of plans and elevations. To supplement these, scrap views, assembly and detail drawings are made. Similarly a solid clay or wooden model can be made manually or a model made out of cut and folded card.

Computer aids can model conventional flat drawings. These models are in the form of multi-sheet drawing data structures consisting of lines, arcs, and other geometrical, symbolic and textual constructions. These constructions include conic sections and smoothly fitted curves of various kinds, together with strings of text characters. The line styles of these graphical constructions, together with text of a mixture of fonts and sizes, can be combined to model any two dimensional drawing within the computer. However, as was indicated earlier in Chapter 4, a number of modelling tools have also evolved to support the modelling of three dimensional objects.

The following types of computer aided modelling techniques are now available and their limitations need to be understood by practising engineers wishing to use them:

- Three dimensional wire frame modeller;
- Three dimensional solid modeller;
- Three dimensional surface modeller.

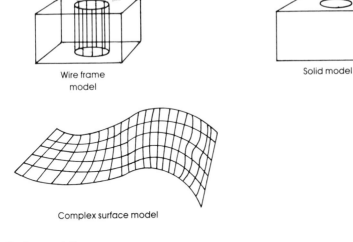

Wire frame
model

Solid model

Complex surface model

Design modelling.

Wire frame modelling

Within a so-called 'wire frame model', the computer models edges and vertices of each of the surfaces of the three dimensional structure of the object. The lines and vertices which form the edges to the various surfaces can include internal surfaces within the object such as a hole passing through it. Regular curved surfaces can be approximated by a ruled surface or a surface of revolution about a declared axis which follows the contours of the surface to be modelled. The key point to understand with such wire frame models is that the model does not contain sufficient information to fully model the solidity of the object. It has no 'knowledge' of what is material and what is void space.

The lines and vertices are modelled using three dimensional co-ordinates to define the ends of the lines, vertices and arc centres to which lines are connected within the various surfaces used to represent the object. Such wire frame modellers enable the user to view the object from any position with different perspective angles and centre lines. The views of the wire frame surfaces can become quite confusing to the user since all of the lines and vertices held within the wire frame model by default are displayed. For such models of even modest complexity, the viewed object on the screen can become a rather cluttered picture of many lines and vertices. Optical illusions can also occur and the user can find difficulty in distinguishing surfaces which are convex and those which are concave.

Having said this, many regular users of wire frame models do adjust to their limitations and are prepared to live with their shortcomings of clarity. Their advantage is that the time taken to compute the wire frame representation is shorter than with other forms of solid model representations.

Algorithms are also provided to remove most of the hidden lines which result from the viewing of the object from a chosen angle. This facility can produce what looks like a true three dimensional 'solid' representation of the object. However the computational load and hence time to create the view of the object with hidden lines removed can become quite high, particularly for objects of complex geometry.

The computation of sectioned three dimensional views of wire frame models can produce some strange results unless the sectioning plane is one of the surfaces which have been implied within the wire frame model. This is because the computer only 'knows' about the edges of surfaces (and the vertices where these edges coincide) which describe the object and not the solid nature of the object.

Solid modelling

Full three dimensional solid modelling techniques have evolved using a number of computational approaches. In the first instance solid representations were developed for primitive volumes such as cubes, spheres, cones, prisms, toroids and frustrums. Such objects can then be combined using logical statement (Boolean expressions) in order to fully model objects which are more complex in shape. Thus, primitive volumes can be logically added and subtracted at nominated orientations and reference datum planes to build up quite complex solid three dimensional models.

This method of modelling solid objects provides a convenient method of approximating many types of components. These modelled components can be further combined to provide complex modelling scenes. A classical example of this type of modelling can be found in applications such as chemical process plant modelling, where major pieces of process plant and piping components such as valves, pumps and piping elbows can be realistically modelled from combining primitive volumes. Such models are a powerful aid to visualising complex process plant and validating that the plant is correctly connected. Further, such problems as clash conditions between piping runs and between piping runs and plant equipment can be detected within the computer and eliminated at the design stage.

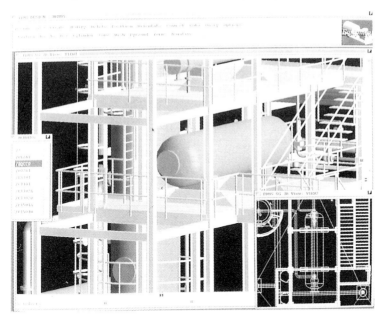

Screen image of a computer generated 3D solid model of a process plant design (courtesy of CADCentre Ltd).

In addition to three dimensional solid modelling, other methods of generating computer aided full solid model representations have been developed. For example, one such powerful method enables the user to extrude

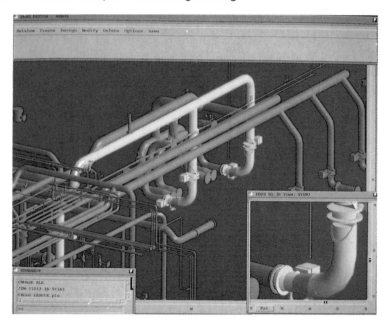

Screen image of an enlarged part of the process plant model (courtesy of CADCentre Ltd).

an area profile, either as a two dimensional plan or elevation, along a vector to interface with another three dimensional plane. This allows quite complex three dimensional volumes to be created using a technique naturally familiar to many engineers. These extruded solid models can be further enhanced by logically combining them with primitive volumes as described earlier or by combining them logically (adding, subtracting or exclusively or-ing) with other similar extruded models.

A full solid model representation allows automatic sections to be taken through using any three dimensional cutting plane. The interface geometry at the cutting plane can be automatically reconstructed to provide accurate visualisations of the sectioned view of the model. Also, since the full solid model 'understands' where material exists and where it does not, it is possible for the volume and other derived parameters to be calculated automatically from the model e.g. centre of gravity or moment of inertia.

Complex surface modelling

Many surfaces which occur in engineering are, by their nature, complex. Where the surface is doubly curved, that is in both directions, special surface modelling tools must be employed. Doubly curved surfaces are found in such

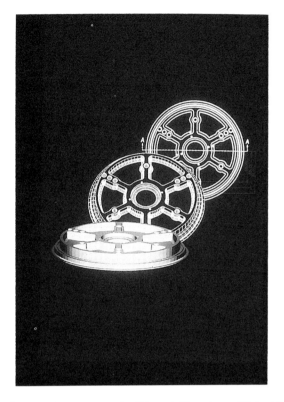

A typical visualisation of a component solid model (courtesy of Robary Ltd).

applications as automative and aerospace engineering, clothing and shoe
design and plastic moulding and cannot be modelled without recourse to
special surface modelling software tools. Complex surface modelling tools
involve structuring the surface as a set of small patches which can be defined
by a variety of mathematical techniques. The total surface model is controlled
using the surface modelling tool parameters to manipulate these patches and
the boundary conditions between them.

The patches used for surface modelling are polygons. They are defined
by the mathematical representation of various types. Examples of patch types
found in surface modelling tools include B-splines, bi-cubic, Bezier, Coon
and NURBS. A good surface modelling tool will support a variety of these
patch types since different types of patch are best suited to different styles of
surface modelling. Clearly, at areas on the surface where the curvature is
changing most rapidly (for example at sharp bends or corners), the density of
patches must be greater if an accurate representation is to be achieved. Also of
importance is the order of the polynomial used to model the surface patch; the

higher the order of the polynomial the greater the potential for overall accuracy of the surface model.

Another approach to surface modelling consists of automatically fitting splined surfaces over cross-sectional profiles (similar to fitting the skin to the hull of a boat). In this case the cross-sectional profiles need to be closer together where the curvature of the surface is high in order to achieve good surface accuracy. This approach is well suited to surfaces of components which are to some degree annular in shape, e.g. manifolds, ducts and modified linear extrusions.

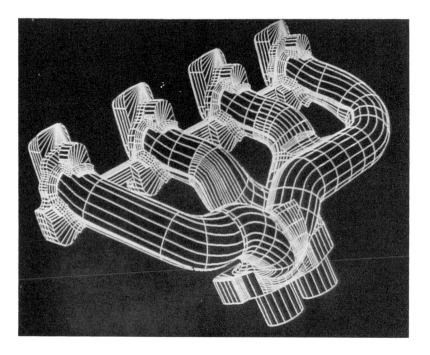

Wireframe surface model of an exhaust manifold (courtesy of Delcam International Ltd).

Surface modelling tools can be combined with either wire frame or full solid modelling tools to provide a comprehensive range of modelling techniques. Such combinations are capable of creating extremely realistic and accurate models of objects and scenes. Their use is to be found in an ever increasing range of engineering applications.

Viewing models with surface properties

Three dimensional models can be further enhanced within the computer by assigning 'material' properties to their surfaces such as texture, reflectivity, colour hue and intensity. Model viewing tools for such techniques are now extremely advanced. Users can position their viewing point, viewing angle, vanishing point and assign scale parameters for the view together with positions of multiple 'light sources'. The multiple light sources can be positioned to enhance the quality and realism of the view of the solid model on the screen. Moreover, these viewing techniques can be further refined to generate a sequence of views which simulate walking around or through a model scene.

Such techniques can be a powerful aid in exploring design alternatives at the conceptual design stages of a product. They also offer a powerful method of communicating designs between technical design staff and non-technical personnel such as marketing and financial staff or indeed financial backers or planners.

For many small and medium sized companies, the potential additional benefits of full three dimensional modelling and visualisation techniques do not justify the extra effort in constructing the necessary computer models in the first place. After all for many hundreds of years engineers have successfully carried out their design, development and communication activities by using orthogonal two dimensional views of the objects and assemblies which they design. The use of three dimensional modelling and viewing tools can be a powerful aid in a number of applications within engineering but there is a danger of management staff being carried away with the appeal of it all. Like many other areas of technology, advanced three dimensional modelling and associated viewing tools need to be carefully assessed against the activities and aspirations of the company. For some companies who are predominantly design houses, the full modelling techniques set out above could be justified on the basis of easing communication with their clients and on prestige grounds.

Having explored the physical appearance of the conceptual design together with the physical outline of the main sub-assemblies and mechanisms, the functional performance of the product must also be considered. Indeed the physical form of the product must be compatible with the constraints placed upon the product by the functional performance consideration.

Conceptual design performance - analysis and simulation

The functionality of the product or system will have a number of aspects. The conceptual design activities will include preliminary simulation and analysis of some, if not all, of the following functional constraints:

— Strength analysis of components, sub-assemblies and assembled product;
— Thermal performance;
— Flow analysis;
— Power analysis;
— Ergonomic performance;
— Outline performance analysis of any mechanisms;
— Outline functional analysis of any system content at the top level of the block diagram (e.g. instrumentation and control system or computation system).

A wide variety of analysis aids are available which are marketed by turnkey system houses, software houses and computer bureaux companies. Vendors tend to focus their offerings around specific engineering market sectors. These analysis and simulation aids can provide confirmation of the basic acceptability of the functional performance of the product or system at the early conceptual design stage of the development process.

A typical example of this type of initial analysis concerns the ergonomic consideration of the new product. It is possible to relate the physical topology of the proposed product design in one or other of the dimensional computer models outlined earlier in this chapter, to a similar simplified model of a standardised human operator. The computer model of the human user or operator can be built up of standard components such as trunk, head, leg and arm segments to form a so-called manikin which is capable of being animated. By changing the angles between the manikin component parts (much as one can simulate the operation of a mechanical mechanism), the locus of movement of its component parts can be displayed.

Variants of such computer model manikins can be made to represent the normal range of distribution of men, women, children and perhaps different racial groups. By displaying selected positions of the manikin models in relation to the computer topological model of the new product in orthogonal views, the ergonomic performance of the new product can be explored and initially optimised. Early consideration of the ergonomics of a design, prior to any later need to build physical models, can provide additional confidence in the design concept. It helps to reduce the risk of later rework with the attendant hazard of delays in bringing the product to the market place.

This example is characteristic of the impact of computer aids on the design process. The emphasis is on investing total effort (including that effort contributed by the computer) at the earlier stages of design. The objective is to explore more design options at the outset and to maximise the chance of going forward to the next stage of the design cycle with lower technical risk. Such an approach can lead to significant improvements to both the quality of design of the final product and greater acceptability of the product by the market place.

Poor aspects of a product design which are built in at the early conceptual design stage can cause problems in manufacture, test or in maintenance. This can arise particularly in medium sized companies because of insufficient or late communication between the design function and the downstream functions of manufacturing, planning, inspection, test and maintenance. Small and medium sized companies tend to grow beyond the stage when informal and unstructured communication can adequately handle the technical issues which arise. This is particularly the case where product complexity tends to increase, or where initial single products develop into ranges of products and product variants which are frequently perceived to be needed to service the company's developing market.

Sharing information and resources

By developing the conceptual design information within a computer environment and by networking the computing facilities with other key functions within the company, the potential exists to make key technical staff within the downstream function aware of new product design concepts at an early stage. Naturally, wider access to such information can have an attendant commercial risk of premature leakage of information to the outside world and particularly to potential competitors. In order to mitigate against this risk, it may be necessary to limit the access to sensitive information on new products to company staff who have been assigned password protected privileges within the company. This may be important in companies engaged in commercially sensitive work.

The ability of these key staff to explore the early computer models of the evolving design concepts of the product offers potential to influence the early stages of design. This potential for improving the communication between the technical functions of the company can help to ensure that new products are designed from the outset with due consideration for the manufacturing, inspection, testing and maintenance processes. The strategic use of the computer environment can help to ensure that each of the technical functions of the company can hope to work in a 'surprise free future'.

General purpose computer aids such as computer aided design and draughting systems packages and modestly priced design analysis aids can be of considerable assistance in exploring design concepts. Such frequently used computer aids are now available at costs which are affordable. Sometimes, particularly for small and medium sized companies, the frequency of use of a computer based design aid is not sufficiently high to justify purchasing a lease to use the software, in-house, within the company. This may be especially true of comprehensive general purpose design analysis aids such as stress analysis, circuit analysis, control system analysis, chemical process simulation and flow or thermal analysis. In some cases the computational load is often high and the hardware required is beyond the financial justification of many small and medium sized companies. On these occasions, some consideration should be given to using such computer aids directly on a computer bureaux service, using appropriately trained company staff. Alternatively, access to the technology can be gained, albeit second-hand, through a specialist computer based design house on a sub-contract basis.

Linking concept design data to downstream functions

Many of the most successful applications of computer aids in the conceptual design stage include facilities to link the design concept data directly through to the downstream detail engineering and manufacturing functions. A simple computer aided sketching facility, with some facilities to embellish the graphical information produced with colour or texture information, can be a relatively low cost initial starting point for using computer aids within the company. It is however important that the geometry model produced by such a computer aided sketching system, can be directly used by a comprehensive design and draughting system which will be needed to work up the sketch information into fully dimensioned and annotated engineering drawings.

Some of the most impressive systems which include provision for conceptual design are to be found within companies which have invested in bespoke development or customisation of their computer aided system. Such classic systems include specialist systems aimed at sharply defined product areas.

An overview of a shoe design computer aid

A good example is a system for the complete design and manufacture of shoes. Within this example the conceptual design of a new range of shoes is absolutely crucial to the success of the new product range and of the shoe manufacturing company. In this case the basic fashion style is captured into the computer by scanning the co-ordinates of a fashion shoe last model for next year's fashion. This style may originate in a fashion centre in say Italy in the late summer. The scanned three dimensional co-ordinates are converted into a doubly curved three dimensional surface model within the computer. The shoe last surface is then trimmed to the required style line of the top opening of the shoe by drawing the style line on to the three dimensional surface model. This is done by using a cursor slaved to a tablet stylus which is moved in two dimensions over a tablet surface but transposed on to the three dimensional computer surface model.

The resulting three dimensional surface of the upper shoe surface is further edited by drawing with the tablet stylus on to the surface model on the screen in order to identify parts which the designer may wish to remove or add to. Stitch pattern and other cutting lines can be added in a similar manner. Standard features of decoration such as bows, buckles and straps can be called up from a library of such features and added to (or removed from) the surface model of the shoe in order to refine the conceptual design. When a candidate design has thus been created on the graphics screen, the resulting composite model of the upper shoe can be checked against certain design measurement criteria which are held within the computer software in order to check and confirm that the shoe conforms to these known measures for quality of fit.

There are also facilities within such design systems to mix colours on a colour palette and to assign both colour and surface properties (e.g. shiny patent leather, suede leather etc) to the upper shoe model. Such surface properties can be rapidly changed within the computer model. By choosing viewing parameters, the model can be viewed from any angle on the graphics screen. The basic model which has been developed at the scale of a single size, (near to the middle of the range of sizes), can then be scaled to each size and width fitting within the desired range. The impact of the size within the range may be assessed in the context of the proposed conceptual design. Accurate solid foam models can be cut at this stage at whatever shoe size required typically using numerical control machining techniques. Precisely the same surface geometry which has been produced by the design stylist is passed to the part programming system used to prepare the part program for cutting the prototype foam model.

This example provides a very powerful tool for rapidly exploring design topology, colour and material finish in the context of a highly fashion conscious market sector. It provides rapid feedback to both the designer and other key staff within the company, allowing many more design options to be explored within the short time available before committing to volume production runs to service next season's market.

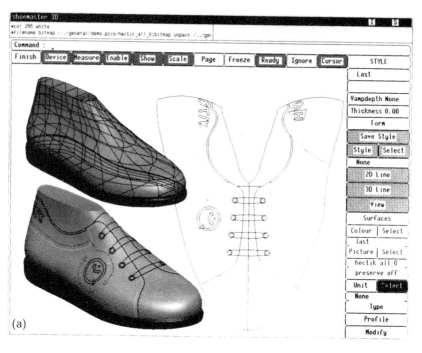

(a)

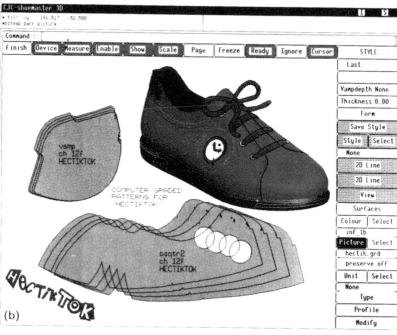

(b)

(c)

Stages of computer aided shoe design: a) Computer generated 3D upper shoe surface models showing initial last surface and part of the shoe design process including stitch patterns; b) Computer model of completed shoe design with scaled flattened components for different sizes; c) Actual finished shoe (courtesy of C&J Clark International Ltd).

What is more, having decided on the shoe designs to be manufactured, there are further facilities to automatically flatten the doubly curved surface of the upper shoe into the required flat surface components. The flattening routine makes allowance for the bending of the material and can also add overlap allowances for bonding and stitching the material together to ensure that the formed surface is accurately matched to the original doubly curved surface model. Having developed the required flattened geometry for each size of shoe within the range, these bounded geometry shapes can be used directly to produce either precisely accurate cutting knives or for driving numerically controlled water jet or laser cutters to be used within the production run.

The above example of the integrated use of computer aided techniques for shoe design and production serves to show the direct advantage of linking the full product development cycle from the original conceptual design through to manufacturing tooling. It shows potential for largely eliminating

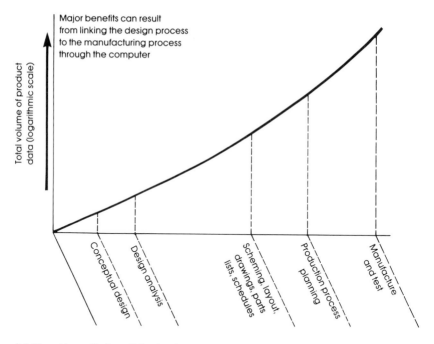

Additional benefits from linked systems.

the conventional design detail draughting and minimising prototype modelling processes. Whilst shoe design is not normally within the capabilities of most small and medium companies, it demonstrates the potential of customising available computer application software into a system closely targeted to a specialist market. It also has the advantage of being easily appreciated by technical staff from any engineering sector. They can relate the strategic concepts to their own companies and products.

Codified designs

Other examples which directly relate to small and medium sized companies include special design systems for heavy fabrications. Examples include process plant items such as welded steel storage tanks or silos, heat exchangers and pressure vessels. Here, the design is invariably controlled by a design code of practice which can be enshrined within the application software.

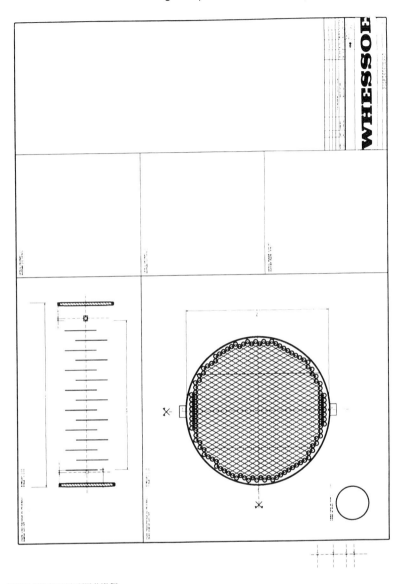

Heat exchanger design: longitudinal and transverse sections of bundle layout (courtesy of Whessoe Computer Systems).

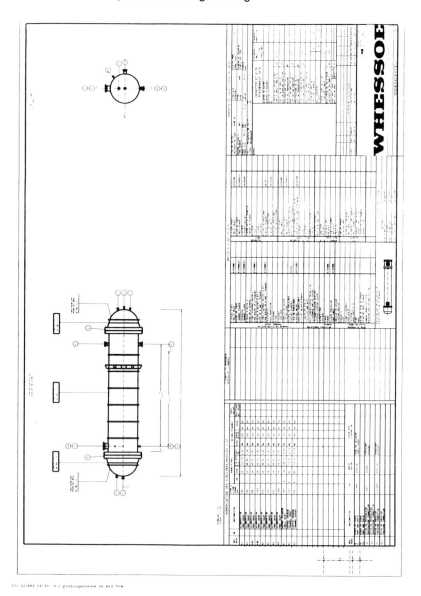

Heat exchanger design: elevation view and design data (courtesy of Whessoe Computer Systems).

Heat exchanger design: finished product (courtesy of Whessoe Computer Systems).

These specialised design aids allow alternative concepts for equipment design to meet particular operating parameters to be rapidly explored. Choices can be made to optimise production methods and costs or to minimise stock holdings.

Conceptual design of equipment packaging

Similar examples are to be found in system design. As was cited earlier, applications where a system is built on a 'pick and mix' basis, using standard components and modules are particularly well suited to being explored at the design concept stage using computer aided techniques. The physical fitting of the system modules into equipment racks and enclosures can benefit. These tend to be built from standard ranges of largely metal or moulded components and can also be accommodated within associated standard computer aids. Appropriate computer aids can be used for exploring the physical topology. Consideration can be more effectively given to packaging aspects (sometimes called space management). Examples include such aspects as alignment and fit, and physical clashes of components and subsystems within the available space. These can be interactively explored at the conceptual design stage and potential downstream problems minimised.

Again, standard ranges of enclosures, panels and racking modules can be established as library items on the computer based system, hence speeding the process of optimising the physical layout concept for the system units. Important aspects such as control panel layout concepts can be explored interactively on the screen. As with the shoe design example, sheet metal enclosure components and panels may be automatically unfolded (developed) by computer software. The specialised automatic unfolding software routines take account of such details as bending allowances for the type and thickness of sheet material specified and any relief holes at corners. They generate the developed geometry of the flat sheet components from which the sheet metal work is to be made. Again this developed geometry can be used directly as input for the numerically controlled punching, nibbling and bending machines.

Automotive and aerospace conceptual design

Within the automotive engineering sector, the development of engine, transmission and body design concepts using computer aids within the major vehicle manufacturers is now universal. However, much of the detailed component and subsystem design and manufacture is often sub-contracted to medium sized companies. Traditionally, there were problems in maintaining the precise accuracy of complex geometry of the overall concept design stage, often expressed as a wood, plaster or wax model. The problem of the so-called 'model maker's thumb' resulted in small changes of geometry as the various sub-contractors worked up their detail design from the drawings and model samples provided by the vehicle builders. When these detail components and subsystems were eventually brought together in final vehicle prototype assembly, there were often problems. Examples include alignment and fitting of components, small discontinuities of surfaces at joining edges of surfaces (e.g. body panels) and asymmetry in body geometry. In a number of vehicles designed before the widespread use of computer aided techniques, right hand doors were of a slightly different curvature geometry to left handed doors, wing panels could have surface discontinuities with door panels, sunshine roofs failed to accurately match the roof line etc. The final geometry of the built vehicle may have strayed significantly from the lines originally laid down by the vehicle stylist's design concept as expressed in the original drawings and models.

As a direct result of building the stylist's design concept as a computer model rather than as sketches and hand made models, the stylist's geometry can be precisely captured in a numerical model form within the computer. There is a declared intention and indeed, for the major vehicle builders, an enacted policy of passing master geometry to their sub-contractors in numerical form. This is extracted from the developed stylist's geometry model rather than in the form of paper drawings and models. This master geometry ensures that the sub-contractors are confined to work to the master vehicle style lines. Thus, for example, the medium sized sub-contractor which designs the sun roof panel and mechanism or the air spoilers components for the sports version of the basic vehicle, is confined to work exactly within the original style line geometry of the vehicle. Similarly, the sub-contractor, making the roof panel tooling, also receives the same numerically defined geometry data for the roof curvature. Thus, both sub-contractors work to the same master surface geometry data, hence maximising the chances of good compatibility of their components.

Solid wax or foam models can still be produced in order to express the design at the conceptual stage in a solid form. These are also derived directly from the original stylist surface models held numerically within the computer. Using numerically controlled machine tools, they can now be cut under the control of precisely the same numerical data as is held in the stylist's surface model - the same data as has been passed to the many specialist sub-contractors. Historically, the model was of necessity, a somewhat subjective interpretation of the relatively ill defined stylist's drawings and sketches by the model maker.

What is more, the successive small variances and differences in interpretation, resulting from successive models and patterns made by each of the pattern makers within each of the sub-contractors can now be eliminated. The company structure of vehicle builders and their associated network of sub-contract medium sized companies (which characterises the automotive industry), can now be constrained to work with the same definitive master geometry for the complete vehicle.

Equivalent structures are to be found in the aerospace industry. Similar advantages from controlling the transfer of master design data throughout the network of sub-contractors can be achieved. The implication of this approach is that for many sub-contractors within these key industries, the use of computer aided techniques is becoming mandatory. They must be in a position to pick up master data in numerical form and pass back to the main contractors their engineering data, including the geometry resulting from their own design activity, in a compatible numerical form. This data can then be fully integrated into the overall product design within the major vehicle or aerospace builders.

Future trends

The advent of knowledge based systems and so-called artificial intelligence software techniques within computer aids will expand the range of specific computer aids aimed at specialist engineering market sectors. Using these emerging techniques, the methodology of the design processes which relate to specialist engineering activities can be increasingly included in the application software. This approach enables additional disciplines to be incorporated into the conceptual design and later design stages. These new techniques could assure conformity to best design practice and increasingly improve design for manufacture and for the test and maintenance aspects of new designs.

7

Computer aided draughting

A number of design concepts need to be explored, refined and discussed with other functions in the company including marketing, production planning, inspection, test and maintenance. Following this process, a consensus decision must be made as to the product or system concept which will form the basis on which detailed design will proceed. The resources required for full scale development to be completed must be assessed. The objective is to produce a full set of documentation from which the new product or system can be made. Often a pre-production prototype will first be made for further evaluation.

Information growth in the product/system cycle

At this point in the product cycle the volume of information relating to the new product starts to grow rapidly. A design concept for a typical product or system within a small to medium sized company may be expressed in a relatively small number of drawings, outline specification documents and initial analysis results. By the time a full set of manufacturing information has been produced this volume may have grown to hundreds of drawings and other associated documents. The maintenance of the relationships and dependencies of subsets of this information are crucial if problems are to be minimised in the manufacturing and testing functions.

Much of this documentation consists of engineering drawings, diagrams and their associated parts and materials lists. The design drawing office has long been recognised as a major potential bottleneck. Computer aids for developing and modifying engineering drawings have been the single most widely used application of computer aids within the engineering community. Computer aided draughting is often the starting point within a company when it seeks to upgrade its technology base. The capital cost of hardware and software required to support basic computer aided draughting has been substantially reduced by the advent of powerful personal computers. Such hardware now provides high speed processors (25 MHz clock speed and greater), maths co-processors, large volumes of hard disk, graphics tablets and high resolution graphics terminals with a range of colour options. The functionality of draughting software which can be supported by such hardware has also matured to the point where a comprehensive set of functions is provided to meet the day to day needs of the engineering draughtsman.

Applicability of computer aids in the drawing office

However, the use of computer aids in the drawing office must still be carefully considered if the technology is to be used to its best advantage. A small to medium sized engineering company needs to assess the applicability of computer aided draughting to the mix of work within its various engineering drawing activities. The effectiveness of the technology is dependent on both the mixture of drawing tasks and the way in which the technology is implemented and customised to meet the company's needs. Many companies have failed to recognise these needs. They have gone ahead with the purchase of computer aided draughting equipment without sufficient planning and training and consequently failed to achieve the benefits which they had sought.

In making an assessment of the applicability of computer aided draughting, it is important to have an understanding of the characteristics of drawings which benefit most from being prepared using such equipment. Drawings rich in these characteristics will benefit most in terms of the speed with which they could be produced by using computer aided draughting. The productivity of the draughtsman using computer aided draughting equipment will therefore vary accordingly.

Drawings well suited to computer aided draughting

What then are these drawing characteristics which are predisposed to the successful application of computer aided draughting techniques? How can a company plan to apply the technology to its mix of draughting and associated activities?

Application libraries

The first and most obvious example of these drawing characteristics concerns the potential to re-use drawing related information held on computer disk based libraries. For detail design processes which contain a high degree of combining standard components, the case for using computer aided draughting is likely to be strong. Standard components are typically expressed within the detail design documentation either as non-dimensioned symbols or dimensioned mechanical type drawings. Electrical or electronic circuit diagrams and logic diagrams contain clear cases of the former type of component as do piping and instrumentation diagrams. Using traditional manual methods of draughting, applique or templating techniques have been used for many years. This involves transferring standard components from sheets of appliques or drawing using a stencil on to appropriate places of the design drawing or diagram. Examples of the latter type of component are less clearly defined in a general sense but process plant and piping drawings, conveyor systems, heating, ventilation ducting and control panel layout designs are clear examples where these types of component are likely to be found.

Any credible system which is offered as a competent computer aided draughting system will normally be supported by a range of compatible libraries of standard components, orientated to particular applications. Such libraries which are appropriate to the broad sweep of manufacturing industries will include the following:

— Electrical circuit symbols;
— Switchgear circuit symbols;
— Electronic logic component symbols;
— Hydraulic system component symbols;
— Piping and instrumentation component symbols;
— Mechanical fixing and fastening component mechanical drawings;
— Piping component mechanical drawings;
— Ducting components.

Clearly, the ability to select a drawing or symbol of a component from a library, closely related to the design application, is a very effective aid to the design draughtsman. This approach should be extended wherever possible by building up further libraries of standard components and sub-assemblies which are specific to the company's operation. The libraries which are so constructed become an important adjunct to any standard computer aided draughting system. If the nature of the detail design work lends itself to this manner of working then libraries of components and sub-assemblies can be frequently used. Management can impose additional discipline in order to increase the ease, and motivation, to use company standard components and methods of best design practice within the company drawing offices.

Replication of geometrical constructions

Allied to the potential for using application libraries, is the degree to which geometrical shapes are replicated within a mechanical drawing. With a computer aided draughting system, a geometrical construction can be set up as a temporary 'grouped item' which can be rapidly replicated within the drawing on the screen. Such replications can also be rotated, mirrored about an axis and scaled to a different size. Drawings which contain a degree of replication of similar geometrical constructions are likely to benefit from the use of computer aided draughting.

Parametric draughting

Within many design activities, ranges of products need to be considered. The range may typically cover different physical sizes of the product. In which case, the manufactured components used are often similar in geometrical shape but of differing dimensions for each product variant within the range. The detail drawings of components which need to be documented form families of parts. If such families of parts form a significant aspect of a company's design activity, a special feature of the computer aided draughting can help to improve drawing office productivity. This is known as 'parametric draughting'. The geometry of a component is defined in the form of a template into the computer, using a special design language. This approach allows selected dimensions and dimensional relationships, within the geometric template for the component, to be specified as named variables, or as an algebraic expression involving other dimensions within the component template. Such parametric draughting facilities ensure that accurate scaled drawings of each member of the family of parts can be rapidly and automatically generated when values are assigned to the variable dimensions at the time the parameterised template is invoked by the system user. Great care has to be taken to ensure that each of the variable dimensions of the geometry is unambiguously defined and the dimensional information is consistent and not over-dimensioned. (The nature of such parametric design facilities will be discussed later in this chapter.)

Complex geometrical constructions and calculation

Another case where the use of computer aided draughting can be of advantage is where the detail design draughting involves extensive geometry calculation, or complex geometrical constructions, in order to establish the necessary detailed geometry on the drawing. Examples can include determining transitional geometry between different geometric features on the drawing, exploring the loci of key points within mechanisms, complex surface definitions, exploring tolerances and potential alignment and fit problems in assembly drawings and other difficult geometry cases. Provided that the proposed computer aided draughting system supports a comprehensive set of geometric constructions, editing, and graphical viewing facilities, these difficult geometric cases can be explored rapidly. Extensive construction geometry can be set up in a so-called 'layer' (similar to an overlay) within the computer drawing model.

The ability to hold within the computer a detailed numerically accurate model of this complex geometry ensures that precisely this same geometry is available for activities which are to be carried out downstream from the drawing office. Such activities include further sub-contract design, manufacturing, inspection and testing processes. This is another example of avoiding the problems of the so-called model (or pattern) maker's thumb which was referred to earlier.

Symmetry

A degree of symmetry is often found within many detail mechanical design processes. Many mechanical components which need manufacturing for which detail drawings are required, are turned parts. Such parts, by their nature, tend to be at least symmetrical about the axis of turning. Also sheet metal components are often characterised by being partly symmetrical about one or more axis. The same can be true for other types of components all of which, if they are to be made in-house or by a sub-contractor, need to be defined on a dimensioned detail drawing.

Where such symmetry is a significant feature within the drawing process, using computer aided draughting techniques, it is necessary to construct only half of the symmetrical aspect of the geometry. This symmetrical geometry can then be replicated in its mirrored form about a user defined axis. The 'MIRROR' command facility of the computer aided draughting system can be used on a group of geometry to complete the symmetrical shape. If the geometry is symmetrical about two axes the replication can be performed about the other axis in a similar way. In this case only a quarter of the geometry needs to be set up in the first instance.

A further aspect of this characteristic includes the case where geometric shapes are rotated about a centre point. A good example would be a pitch circle for flange mounting bolts. By using computer aided draughting techniques it is only necessary to define the geometry set to be rotated once and then to invoke a system command to replicate the set a declared number of times about a nominated centre of rotation. This facility also provides for optionally rotating the geometry itself on a separate centre of rotation as it is replicated around the first centre.

Such facilities can allow what appear to be quite complex drawings, containing a high degree of symmetry and rotational replication, to be generated rapidly within the computer drawing model. They also provide for rigorous numerical accuracy in the geometric information relating to the drawing.

Techniques for dimensioning

The dimensioning of mechanical drawings is often seen by the draughtsman as a somewhat tedious and boring activity. The quality of dimensioning information and its consistency with national standard codes of practice is highly dependent on the draughtsmen and their experience and attention to detail. Drawings which contain a large amount of dimensioning are likely to benefit from computer aided draughting techniques. A competent computer aided draughting system offers a wide selection of semi-automatic dimensioning facilities. Such facilities provide for meeting different national dimensioning standards by user definable options for selecting the dimension annotation style.

The normal technique for dimensioning, using a computer aided draughting system is flexible and quickly learned. The user points sequentially (using the screen cursor) at two features on the drawing such as an intersection of lines, a centre of an arc or circle, an individual line or an end of a line and then indicates where the dimension is to be shown on the drawing. The computer calculates the accurate length for the dimension, or the angle between the two features, from the model of the drawing. The dimension leader lines, dimension line and arrow heads are automatically generated together with the actual dimension in the format which has been elected by the user's system parameter choices. Dimension types are provided for horizontal, vertical, true length, angular etc dimensioning. Chain dimensioning from a nominated datum line or point can be particularly rapid.

The ability to zoom into sections of the drawing and thus to enlarge drawing features, using conventional graphics windowing techniques, allows the precise feature on the drawing to be identified and confirmed to the draughtsman during the dimensioning activities. Additional textual information, such as tolerance information, or machining symbols can also be added to the semi-automatically generated dimension information. The text size and style for dimensioning activities can be set by the user and thus optimised for ease of reading.

The use of a flexible set of dimensioning facilities within a computer aided draughting system can help to eliminate much of the tedium of manual dimensioning. The consistent quality of such information as dimension lines, leader lines, dimensioning arrow heads and text quality can be assured. That is not to say that such facilities will make a poor manual draughtsman into a good computer aided one. The dimensioning process has always relied on a draughtsman having a good feel for a clearly presented drawing layout and the ability to dimension work in a manner appropriate to the manufacturing and inspection processes. Computer aided draughting, particularly when dimensioning the drawing, does not eliminate the need for these skills in the draughtsman.

Additional graphical information

Some drawings require extensive use of additional graphical information in order to clarify the drawing or to meet drawing conventions associated with different industrial sectors. Such drawing adornments include hatching and shading of a variety of styles, various line styles for different types of information, adding indications of texture for different kinds of surface, machining finish symbols etc. Many of these features can be time consuming to achieve in manual draughting. The time spent embellishing the drawing can adversely impact on overall timescales for the product - even if in some engineering applications custom demands that such embellishments are expected.

If the mix of drawings within the company requires extensive use of such embellishments, advantages will accrue from the use of a computer aided draughting system. The system enables the user to quickly define bounded areas within their drawing to be filled with one of a number of shading or hatching patterns. A set of standard patterns is provided within a library or they can be specially defined and used by the system user. Line styles of various types and machining and surface finish symbols are similarly available from a library of such features within the system.

Whilst care must be taken by the user in selecting the areas/features of the drawing for these graphical embellishments, the execution of the additional drawing information is rapid (provided that the system is not under resourced by its hardware) and the consistency of the result is again assured. Once more, there is no substitute for the draughtsman's skill and experience in selecting the features of the drawing to embellish and the manner of the embellishment, but the desired result can be achieved more speedily.

Textual information

Closely allied to this aspect of drawing is a consideration of the amount of textual information which is included on the drawings produced by a company. The text may be directly associated with such graphical features as component symbols (as in the case of circuit or logic diagrams), e.g. circuit component references, component types, part numbers, connection terminal references, values and tolerances. This type of textual information forms an essential part of the circuit information. Without it the diagram would have little meaning with regard to building the circuit. Alternatively, the text may consist of extensive notes to assist in the manufacturing or testing process. Such manufacturing or inspection notes may consist of standardised sentences or paragraphs which are selected in combinations as being appropriate to the particular detail or assembly drawing.

Manually adding such text and annotation to a drawing can be a boring and often repetitive task. In the case of a circuit diagram or other forms of a symbolic schematic diagram, the text is crucial to the completeness and integrity of the engineering information held on the drawing. Missing text within such diagrams can cause confusion and delays later in the component purchasing, manufacturing and assembly processes. In this case the text is intimately associated with the graphical symbols. Within a competent computer aided draughting system such textual attributes of the symbol are directly linked within the drawing structure model. Since for such schematic diagrams, the need for these relationships are known, for each type of symbol used, the availability of the information can be validated within the full drawing model. Hence missing or unassigned component types or connection pin reference numbers can be detected during an automatic validation of the diagram, prior to release of the diagram to downstream functions. These functions may include the design and production of a printed circuit board or the manufacturing planning and materials control functions.

In the case of annotation which is added for the guidance of the manufacturing planning and inspection or testing functions, standard notes can be called up from a text library within the computer aided draughting system or by direct entry from the computer keyboard.

In either of the cases of textual annotation, the size and style of the text can be selected by the user of the system. As with the case of dimensioning text, it can be optimised for ease of fitting textual information into the drawing space available and for ease of reading. Such text can be entered more rapidly from the computer keyboard than manually entered text. It will also be of a consistent quality and to an agreed style, irrespective of the draughtsman. The text can be positioned optimally on the drawing or diagram on the graphics screen using the cursor slaved to the tablet pointing device or a mouse type screen cursor control.

Scrap/sectional views

For the purposes of clarity, it is often necessary to show scrap and/or sectional views on mechanical drawings. If there is strong requirement for drawings with such additional views, computer aided draughting technology could be of assistance. The ability to zoom into a feature, or a particular area within the drawing, and to copy and scale up the view of this area enables scrap views to be automatically created from the main plan or elevation view on the drawing sheet. In the case of sectional views, it is possible to copy dimensioned geometric features from the plan or elevations on to the section view. This speeds their creation and retains the dimensional accuracy of sectional views, provided that they are orthogonal to the axis of the main drawing views. It is helpful to be able to extend lines which define the cutting plane in order to help determine the geometry of other sectional features.

If extensive use of sectional views is required within the work mix of the drawing office, serious consideration should be given to using a computer aid which supports some form of three dimensional modelling (either wire frame modelling or full solid modelling). The automatic creation of sectional views directly from the three dimensional computer model, for any nominated cutting plane, can provide an extremely beneficial aspect of computer aided modelling. The generation of sectional views of complex doubly curved surfaces is another case where the surface must be modelled by complex surface application software.

It is true that to set up a three dimensional model within the computer can be more time consuming than modelling the part or assembly as a series of two dimensional drawing view models. However, the more rapid production of sectional views (as well as other views, e.g. three dimensional) is likely to justify the effort of using three dimensional modelling techniques. Much will depend on the frequency of the need for these additional views. The decision to use such modelling techniques can only be made after an assessment of the mix of drawings typically required by the company has been made.

The inclusion of comprehensive three dimensional modelling software will, however, impose a greater computational load on the computer hardware than two dimensional computer aided draughting software. In order to reduce the risk of incurring excessively long system response times it is probable that at least a faster processor will be required, particularly if the complexity of the objects to be modelled are other than trivial.

The combination of the drawing features, within the company's work, which have been discussed so far within this chapter will have a considerable impact on the applicability of computer aided draughting and computer aided three dimensional modelling technology. Any company which is considering the wider use of computer aided draughting must attempt to make a formal assessment of the mixture of drawings which are typically flowing through the drawing offices. If a company finds that its normal or projected work load contains a significant content of a mixture of these characteristics it is likely that some direct advantages will be gained from the introduction or extension of computer aided draughting. The higher the content of these drawing characteristics the greater will be the potential for improving the direct labour productivity within the drawing offices which are associated with the initial production of the engineering drawings and diagrams.

Having captured the engineering drawings within a computer aided draughting or modelling system, these drawing files, held on the computer disk unit(s), are available for both inclusion in other designs or for further modification and additions. It is quite common that an engineering drawing will undergo a number of revisions and re-issues throughout its lifetime. Typically, an average engineering drawing will go through four or five revisions within many companies during its commercial lifetime.

Editing drawings

Using manual methods, the revision and amendment of engineering drawings can be time consuming and error prone. If the required amendment will not easily fit into the available space on the drawing, either the additional information becomes difficult to interpret, because it is crammed in, or major retracing of the original drawing must be made. This process again can be a source of other errors.

Extensive graphical editing facilities are provided within computer aided draughting systems. These enable changes, even major layout changes, to be made much more rapidly than manual methods allow. It is true that the revised drawing must then be re-plotted on to drawing film and for large and complex drawings this can impose an additional time penalty. The overall savings in time to achieve high quality revisions of drawings and diagrams is often the single greatest direct saving in man hours within the drawing office. Also, these savings will apply to all drawings which have been captured into the system, irrespective of their initial suitability in respect of the drawing characteristics which they contain.

Potential gain in productivity

It is clear that potential gains in the productivity of drawing office staff from the deployment of computer aided draughting and modelling technology, are highly sensitive to the mix of work within that drawing office. The vendors of such systems tend to be over simplistic in their statements of such potential gains in order to attempt to justify the capital investment in their products.

As a general guide, the potential gains in productivity ratio in staff time directly associated with initially producing symbolic schematic diagrams using computer aided draughting is likely to be between 2.3:1 to 4:1 and other similar 'pick and mix' type drawings and diagrams between say 2:1 to 3.5:1. Similarly, for mechanical detail and assembly drawings rich in the characteristics described earlier, this ratio is more likely to be 1.5:1 to 2.5:1. For other drawings the technology may show little if any direct benefit during the initial production of the drawing. Indeed for some small and simple drawings it is probable that the overhead set up time for using computer aids will be longer. However, for the editing and amendment of drawings the equivalent ratio is likely to be in the range 2:1 to 4:1.

Care must be taken when using these guideline ratios since they relate to the actual time spent in the drawing and documenting activities. When all of the activities of drawing office staff are monitored and analysed, as was hinted at in Chapter 5, it is invariably found that the actual time spent on the drawing tasks themselves are reduced by other tasks such as:

— Searching out information from other drawings and catalogues;
— Discussions with other design staff;
— Discussions with other functions within the company (manufacturing, inspection and test, sales etc);
— Design and geometry calculations;
— Checking of drawings and other documents;
— Preparing parts and material lists, connectivity schedules etc;
— General administration.

Often for the more senior drawing office staff, such as section leaders, the actual time spent in getting lines and text on to paper in the drawing process can be as low as about 25% to 40% of their time. Even for an average grade draughtsman this time rarely exceeds about 60% of their time, whilst for more junior staff the figure might rise to around 75%. By combining these various guideline figures, the overall direct saving of total drawing office labour time from the use of computer aided draughting technology (over normal manual draughting methods) can typically range between about 15% to 30% for many experienced staff. This figure rises to about 60% for more junior staff engaged in activities well suited to the technology.

If computer aided draughting technology appears to be appropriate to a company, the first step is to consider the functional facilities of a system which will meet its needs. In making an assessment of different vendors' systems it is helpful to have documented the functional specification of the system which will be required for the work of the company. The key issues which should be included within this specification are set out in the following paragraphs. There are some basic facilities which are essential for efficient use of a computer aided draughting system and these will be discussed first.

Functional requirements of a computer aided draughting system

A rich and comprehensive set of command facilities must be provided in order to create basic geometric elements. These elements include lines, circles, arcs, conic sections (ellipse, parabola, hyperbola), polygons, pitch circles and spline curves. The basic geometric constructions will need to be combined in order to construct complex two dimensional geometry. Associated with these geometrical constructions, a user definable background grid facility is also needed. This allows geometry to be optionally related to a background grid reference which can be switched on or off by the user. The grid resolution should be capable of being set by the user of the system.

Geometry manipulation

The ability of the user to be able to 'group' geometry elements and objects in a flexible manner for further manipulation purposes will be particularly important. Extensive geometry manipulation facilities should be provided, including mirror, rotate about any axis, move, copy and scale. These facilities should operate on groups of geometry selected by the user. They must allow rapid modification of user specified design sketches or complex detail drawings. The manipulations of groups of geometry will impact on the effectiveness of the user to rapidly build up drawings which contain symmetry or replication of geometry within the drawing. They are also very important for many of the tasks associated with the efficient editing and modification of drawings.

Extending and trimming lines

A set of commands for extending and trimming lines, using a variety of conditions relating to other geometry features on the drawing must be provided. These commands facilitate the rapid creation of both construction lines and the bounded geometry needed to define the design features from the construction lines. Similarly, automatic commands should be provided for constructing internal and external fillet curves of user defined radius and chamfers of defined angles or distanced from vertices.

Dimension information

A comprehensive range of semi-automatic dimensioning aids are required. It should be possible to place dimensions on the drawing to appropriate UK, Continental and US drawing conventions and codes of practice. The system vendor should be asked to state the international standards supported by their system. Similarly, the system should provide a range of dimensional units from which the user may select. Examples include, metric, imperial and architectural units. It should be possible to convert automatically between these units within the system.

It is important that the dimension information, created by the system as a result of the user invoking dimensioning commands, should be associated with the geometry within the computer model of the drawing. This is an important point; when the geometry is modified, the dimensions previously created should be automatically updated where logically possible. Failure to meet this requirement could result in the need to undertake extensive re-dimensioning work on the drawing following a modest modification to the geometry. There should also be facilities for handling tolerance information within the data relating to dimensions.

Automatic cross hatching

Automatic cross hatching commands should be provided to assist in the hatching of closed boundaries indicated on the drawing by the user, using a window box selection. These hatched areas should be quick and easy to produce, including automatic avoidance of islands. Users should be able to clear similarly selected hatched areas on the drawing. An extensive range of hatching patterns should be provided as standard together with easy to use facilities for users to specify their own hatching patterns. As with dimension-ing, the cross hatching should be associated with its bounding geometry within the drawing model. If the bounded geometry is moved or its dimensions (although not its topological shape) are changed, the hatching should be re-computed automatically. It should not be necessary for the user to have to re-hatch the modified areas except where the shape of the bounding geometry has been changed.

Flexible layering system

Provision should be made within the system to hold different sets of information relating to the drawing model within different so-called 'layers' of the drawing model. These layers are like transparent overlays which can be combined in various combinations to include subsets or all of the drawing model. This allows, for example, construction lines not to be plotted (with their layer switched off) but they can be retained for possible future use in editing. If the layer on which they are held is switched on, they appear on the screen to assist in subsequently evolving or editing the drawing. Similarly, dimensions could be held on a different layer and hence could be shown (or not) either on the screen or on plotted output. A flexible layering system allows arbitrary sets of drawing data to be placed on chosen layers under user control. It should be possible to select any combination of layers for plotting or screen display.

Window, pan and zoom facilities

Window, pan and zoom facilities must be quick to operate. This allows the user to handle large and complex drawings effectively by moving their working screen area(s) over the drawing area. This becomes essential when handling say size A1 drawings using a graphics screen of say 20 inches diagonal size. Also, multiple active screen windows should be available to permit details within drawings to be viewed alongside the whole or other relevant part of the drawing. The fact that each window can be active allows interactive screen commands to be activated from within any of the displayed windows on the screen. It should also be possible to access other applications from within the draughting system without having to leave the computer aided draughting application.

Parametric functions

The facilities for setting up and manipulating parameterised families of parts may be particularly important for some company applications. If parametric functions are required then they should be powerful. These facilities will influence effectiveness of the system significantly. There are two basic approaches to expediting parametric geometry. In one case the parametric geometry is defined using a type of geometry definition language approach. Using this method, draughting system function commands, with dimension variable names, can be strung together with additional control commands. These are set up as routines (macros) and can be filed. The assignment of values to nominated geometric variables is made at the time that the parametric geometry macro is invoked for use.

This geometry definition method relies on the user building up programs and macros in order to define the parametric shapes. Sometimes draughtsmen find problems using this language type approach. An alternative approach has evolved in which the variable dimensions for the parametric geometry are added in an interactive manner to the outline geometry which is

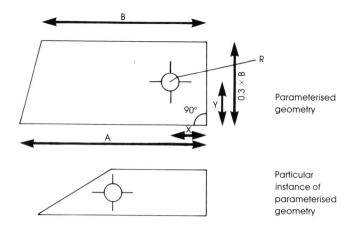

Parameterised geometry.

shown graphically on the screen. From these essentially graphical interactions, the parametric geometry language can be directly generated by the system and the topology of the geometry remains visible on the screen as the parametric shape is being developed. This second approach is often more naturally accepted by draughtsmen. Care must be exercised to avoid ambiguities or duplication in parametric variable definition.

In either method for handling parametrics, the system should provide efficient facilities for setting up geometry macros and must be capable of handling complex parts. Also, the parametrics system should provide support for building logical decisions into the parametric geometry definitions. For example, conditions in the form 'IF (some condition) THEN (some action) ELSE (some other action)' should be supported in the geometry definition. The key point is that the system for programming such parametric geometry definitions and macros must be suitable for use by draughtsmen as distinct from computer programmers. Increasingly, interactive graphical geometric parametric definition is preferred to a parametric language approach.

Parts listing extraction routine

A flexible parts listing (and where appropriate connectivity schedule) extraction routine should be provided. Part numbers and text (or component connection point references and connection links) should be capable of being associated with corresponding drawing elements within the drawing model. The format of the output of the parts and material list or connectivity schedule documentation should also be entirely under user control. Files, containing this extracted information, should be capable of being interchanged with other company systems such as word processors, spread-sheets and databases.

Plotting

Plotting facilities are required for efficient production of the mix of all drawing sizes used by the company. Consideration should also be given to the use of preprinted drawing film sheets. Standard scaled drawing frame macros should be supported for the same range of drawing sizes. Insertion of standard or non-standard notes in boxed areas within standard drawing frames should be quick and easy.

Facilities for plot spooling should be sought in order to release maximum computing resources for the highly interactive draughting tasks. A number of engineering workstations, terminals or personal computers will typically be serviced by one or more plotters, possibly of different sizes and/or spatial resolutions. If plot spooling is not provided it would be necessary to tie up a workstation during the time that a drawing is being plotted. Using plot spooling techniques, drawings ready for plotting can be sent to a special plotter spooling file system to form a queue for plotting. Having sent a drawing for plotting, the workstation, terminal or personal computer becomes free for further interactive computer aided draughting work. The plotter spooling routines service the plot queue and will use any spare computer processor capacity in order to plot the drawings without conflicting with the highly interactive work of the engineering user. The priority of spool plotting must be capable of being varied under password protected control.

Availability of standard comprehensive libraries

The availability of standard comprehensive libraries of components which are appropriate to the work carried on at the company will be an important factor in choosing a computer aided draughting system. Most systems offer some support for basic libraries but often they are either not particularly extensive or not fully compatible with the methods of presentation which are suitable for all companies. Often, for the more widely used systems, third party organisations have developed extensive libraries of different sets of components, both symbolic and mechanical, to meet the needs of specialist engineering sectors. These libraries are either marketed via the main system vendor or directly by the developers of the libraries.

In some cases, as was discussed earlier, there may be a requirement for the system to be capable of being operated as a network with shared access to drawing libraries. Some of the information held within this shared structure will need to be protected from unauthorised access or modification. System vendors should be asked to define the nature of the security and control facilities offered for purposes of file security management. Different levels of password protection are needed to ensure that only appropriate staff may, for example, update standard libraries of components or drawings, authorise the release of product or project drawings and set up access for bona fide system users for a particular product or project.

Handling the computer data

Locating computer held drawings and other documents from a mass of unstructured information can quickly become difficult. The computer aided draughting system should be supported by a computer based drawing registry system in which the relationships and status of drawings and schedules can be modelled. The company must determine how it is to operate this registry system in order to keep track of the drawing and model files which will evolve within the computer aided system. Master drawings are normally kept in a nominated structure within the file directory system in a manner which relates the information to product or projects. Good facilities to hold computer based information in a well structured and appropriately protected manner are essential for the rapid location of a desired drawing or model file. The volume of such information will grow rapidly within a company and its value will soon outweigh the capital cost of the computer aided system. At least several hundred megabytes of disk and back-up tape will be consumed in holding even a modest sized company's computer based drawings and other documents within a very few years.

Decisions will need to be made on how to control the issue, archiving and retrieval of drawings on computer file in a multi-user, multi-product, multi-project environment. The system must be controlled and administered by experienced and trained staff if serious problems of data security and system integrity are not be experienced. Consequently a system administrator must be identified within the user company who must become familiar with both the operating system and application software in order to provide an adequate level of administration service for the system.

User groups

Another key consideration is the availability of an active user group for the system. The notion of user groups has provided a way of sharing experiences amongst the user community for a system. The user group can also act as a powerful pressure group on the owner or developer of the system's application software. It can act to focus the further development of the system towards meeting the needs of the users of the system. Such groups are normally organised by a committee drawn from different types of system user. They hold regular meetings throughout the year at which representatives of the vendor attend. The user group can also be a great help to the developer of the system. It can be used to acquire direct market feedback and can serve as a channel of communication to the users concerning future development plans. Its importance should not be underestimated by a prospective purchaser of a system.

Vendor support

The support arrangement for the system must be clear and comprehensive. The failure of the system when fully integrated into the company's operations can result in major difficulties and delays. Many companies find it advantageous to have a common source of support for both hardware and application software. This arrangement avoids the potential problem of delays through disagreements as to whether the source of any problem encountered is due to hardware or software aspects. Whilst separate supply of hardware and software may result in a lower initial capital cost for the system, this advantage may rapidly disappear if the support arrangements fall down because of arguments over the nature of a system failure during its later operational use.

System response times

As was discussed earlier, a crucial consideration of any highly interactive computer system is the speed of response of the system to its user. The consistency of this response time is of at least equal importance. Target response times for all interactive functions should not exceed two seconds for

95% of executions of the function. The screen repaint times should not be greater than ten seconds for say an A1 sized drawing of typical complexity and textual mix for the company. This guide to interactive response speed must of course relate to the system operation when it is typically loaded for the proposed number of users.

Drawing data transfer

For many companies, particularly those concerned with working as or with sub-contract organisations, the ability to transfer drawing information in magnetically stored form, from and to other computer aided draughting systems will be especially important. It will therefore be essential that any candidate system must offer comprehensive transfer links to probable target systems. Either the system must provide for such transfers through some form of standard neutral format (discussed later in Chapter 13) or via dedicated linking software utilities which are fully compatible with the target systems. In many engineering applications, this can often be a key determining factor in deciding on the system vendor, since certain groups of vendors tend to predominate in different engineering sectors within particular geographic territories.

8

Computer aided process planning

The link between the design drawing office and manufacturing planning function is of crucial importance. Historically, particularly within the UK for too many companies there has been a tendency for the interface to the production function to be less than satisfactory. It is quite often somewhat like a high brick wall with a circular hole in it. The hole is positioned slightly above head height and its diameter is slightly larger than a rolled up drawing! Most communication from the design drawing office consists of posting drawings, associated parts and material list, schedules and other documents through the hole to production planning. Again, for such companies, the return communication from production to the drawing office is rather like hurling brickbats back over the top of the wall! In part, this approach to bridging the gap between drawing office and production has its roots in cultural views of the clear divide between those who 'think' (designers) and those who 'do' (production). Too often within a company, production staff are faced with redesign tasks so that a new product can be made in an effective manner.

This distinction comes about because of the way in which different staff within the engineering design/draughting and production functions perceive the new product, and its sub-assemblies and components. The design and draughting staff have historically set down their ideas in the form of a series of

Poor communication between design and production planning functions.

graphical representations and associated textual documents. They tend to think of the physical appearance and operation of the product in terms of the design geometry, textures, colours, performance graphs, etc. It is their main task to optimise these aspects of the design and to set them down in a series of documents. Production staff however perceive the new product in terms of how it will be made, its structure in terms of sub-assemblies and parts, the methods of manufacture and assembly. In particular, production staff view the components and assemblies in terms of the 'features' which have to be machined, fabricated or manufactured in some way.

From the manufacturing point of view, features have to be progressively manufactured into the parts which come together to form the product. It is the interaction between these features which define the geometry of the component or sub-assembly and eventually the final geometry aspects of the finished product. Thus, if one considers a component which is predominantly made by a series of operations in a machine tool such as a lathe, the turned part can be expressed in terms of features such as:

—   External cylinders;
—   Through axial holes;
—   Internal grooves;
—   Chamfers;
—   Fillet radii;
—   Countersinks.

Different features can be achieved by different machining or other manufacturing operations. The use of features offers a way of defining a component; the various features being associated with different geometrical shapes. For example, a drilled through hole of a known diameter is equivalent to removing a cylinder of that diameter from the part. Features can thus be grouped together and associated with a common type of shape or geometrical entity and hence with a range of alternative manufacturing methods.

The sequence in which features are created within the manufacturing process will also influence such aspects as the quality of the finish of the part, and hence the final product. Clearly, it is preferable to drill such features as cross holes (possibly for lubrication purposes) before an internal hole is drilled or bored, say for housing a bearing. This sequence of machining will reduce the risk of internal burs being formed.

Thus, the production staff are concerned with optimising both the manufacturing method for producing the features of a component and the sequence in which they are manufactured. Best practice rules for such manufacturing methods have evolved and are used in order to augment the knowledge of the production staff. The expression of the component design in terms of features and the sequence of achieving these features provides an alternative high level method of defining the part. The information needed to decide on the most appropriate features to manufacture, the material from which the part is to be made and the required quality of finish can of course be implied within the design information documentation. It is the task of the process planning staff to extract this information from the design documentation. They must specify the sequence of operations which will be needed to incorporate the required features into the component or assembly with the appropriate quality standard.

As a part of this procedure, consideration must also be given to the machines and associated tools available within the company. In the case of machined parts the following information, relating to the available machine tools, needs to be considered in formulating a process plan for a part or sub-assembly:

— The physical limits of the workpiece to be machined;
— The various operating parameters of the machine tool relating to the material of the workpiece;
— The current physical condition of the machine tool;
— The tools which are available and their condition;
— The skills of the available workshop staff.

The process planning staff will use this information together with the features information, extracted from the design documentation, in order to produce the process plan for a component or sub-assembly. The resulting process planning documents list each operation required at each work centre within the production facilities. Process plans contain the appropriate setting up and operating target times for each operation.

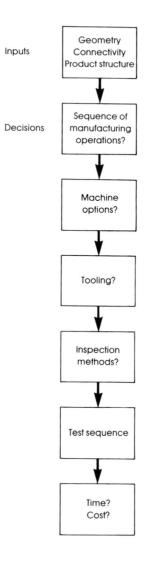

Process planning.

Often, within many products, some components or sub-assemblies are similar, in terms of the operations which are required, to a component or assembly which has been made previously within the company. In such cases, the process planning document for the new part can be based on an earlier one and becomes a variant of it. An additional task in such cases is to identify when this is the case and to gain rapid access to the earlier version.

It is clear from the above discussion that the work of the process planning function is a combination of storing and retrieving a range of standard information relating to the machine tools and their associated tooling, extraction of information from the design data and re-using earlier process plans. Since the design information is increasingly being developed within a computer based system, there is potential benefit from developing computer aids to extract appropriate information for the purposes of generating process planning documentation. The process planning environment has been recognised as being appropriate to computer aided techniques.

The task of holding structured information on machine tools, tooling, and standard set up and operating times pertaining to parameterised features can be handled by standard relational database techniques. The key to the success of using computer aids in this area concerns the approach which is taken in interpreting the graphical information, provided by the drawing office, as a series of features. The process planners can optimise the manufacture of these features using their knowledge and experience. The aim is to use the computer methods and associated communication through it, to break down the wall between the drawing office and the production function, or at least to widen the hole in the wall to a much greater diameter.

Alternative computer aided process planning strategies

There are basically three general approaches to the identification of features which have been used within computer aided process planning systems. They can be summarised as follows:

- Operator assisted feature definition;
- Design by using features;
- Automatic feature recognition.

Operator assisted feature definition

If the design data is held within a computer solid model, the process planner can interactively pick and group parts of the model into features which can be machined and recognisable as such by the computer aided process planning (CAPP) system. These features can then be reasoned about within the process planning synthesis routines. This reasoning process is performed at the physical (manufacturing) level rather than at the graphical level of the engineering model or drawing. Similarly, if the design data is held as a series of two dimensional drawing models, drawing elements can be combined by the user of the system to form physical features.

Design by using features

The second approach imposes a rather new method of thinking about the design of the product and its component parts. In this case the definition of each of the component parts is done by directly specifying the sequence of machined or fabricated features which must be performed in order to evolve the geometry of the part. The features are input, often using a textual language which invokes each feature together with its associated physical parameters (e.g. diameter and length). This imposes a new discipline on the designer since he is forced to think in terms of how the component is to be made from the outset. The resulting geometry evolves by combining features and adding, subtracting, or re-orientating the volumes of material associated with each feature. Generic features are held in libraries which can be customised by the user. These library structures include the facility for property inheritance by means of 'parent-child' links and rules about adjacency relationships. It is a similar process to that which was described earlier in building three dimensional models from primitive basic volumes. This approach can be powerful for most turned parts, milled components and folded sheet material parts. It has the advantage of building 'design for manufacture' into the design from the outset.

Automatic feature recognition

The third approach is based on the CAPP system interrogating the design data, passed from a computer aided draughting or modelling system, and automatically matching combinations of geometry within groups against patterns which are known to identify manufacturing features. This process involves standard manufacturing features being set up as logical models of their associated geometry structures. The interrogation of the design model relies on following sets of rules which relate to classified geometric entities, relationships between adjacent geometric entities and the matching of patterns. The early attempts in this approach relied on a standard set of rules which were included within the CAPP system. More recently, AI technology is being incorporated into such CAPP systems. This AI approach also supports the ability to acquire new knowledge of new rules and feature data structure patterns.

Research is also under way to include machine vision technology to automatically scan hard copy drawings and to automatically build component feature models with which the process planning engineer can interact.

Notwithstanding the different approaches outlined above, CAPP systems fall into two major classes known as 'variant' and 'generative'.

Approaches to computer aided process planning.

Variant type computer aided process planning systems

The variant class of CAPP systems relies on the fact that very little is new and unique within the age old discipline of mechanical engineering production methods. There are a rich range of manufacturing methods. This stretches from various types of metal cutting, shaping and removing (e.g. turning, milling and grinding), through forging, moulding, fabricating/joining (e.g. welding or bonding) to sheet material punching, nibbling and bending. Some features may alternatively be produced by more than one manufacturing method (e.g. circular holes may be punched, drilled or included within a casting). However, for a given quality of finish requirement and range of sizes and workpiece material, this range is reduced to those which represent best practice in manufacture. Each of these preferred methods of achieving a feature can be set up within the variant class of CAPP system as a generic process plan, with variable parameters to be set later in the detail planning process.

Within such a variant CAPP system a 'group technology code' can be used to select an appropriate generic process plan from a structured relational database of individual or families of standardised process plans. These retrieved standard process plans can then be customised by the system user, using the parameters which are appropriate to the case represented by the component manufacturing process. The parameters are expressed in terms of workpiece feature dimensions, feed rate, spindle speed etc. These customised process plans can then be combined to produce the variant process plan for the particular component or assembly. If no appropriate standard process plan exists on the system it is necessary for an experienced process planner to evolve a new standard planning model and to incorporate it into the CAPP system's database.

In order to use such a variant class of CAPP system, the company must first study and classify its manufacturing activities and the families of parts it uses within its product ranges. This task, in itself, can be particularly revealing to the technical management of the company. It forces the company to acknowledge the diversity of detail design and associated manufacturing tasks implied by its range of work. The approach is well suited to companies who have been able to rationalise their product ranges to enable re-use of components of similar geometrical shapes across their range of products. It becomes difficult if not impractical however where there are a wide range of parts of widely dissimilar shape.

The variant class of CAPP systems affords a method of standardising on the manufacturing processes where appropriate. This serves as a method of passing good manufacturing practice on to less experienced process planners. However such systems, in themselves, do not capture the real knowledge or flexible rule structures implied by the skill of the experienced process planner. They merely record a series of instances of presumed skill for each manufacturing sub-process. There is a clear danger of automatically repeating poor manufacturing practice if the standard process plans held within the CAPP system database are in part incorrect or sub-optimal, although there are provisions made for updating and improving this information.

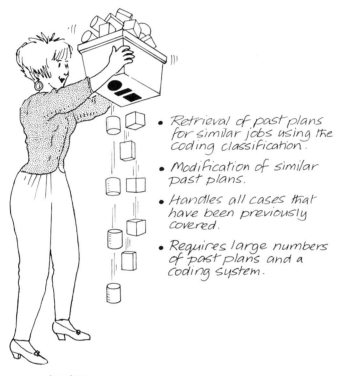

- *Retrieval of past plans for similar jobs using the coding classification.*
- *Modification of similar past plans.*
- *Handles all cases that have been previously covered.*
- *Requires large numbers of past plans and a coding system.*

Variant process planning.

Generative computer aided process planning systems

The so-called generative class of CAPP systems is still relatively new and perhaps less mature. In this case, the process plan is synthesised by constructing plans directly from interrogating the input data in the form of component design data and its raw material specification. The manufacturing processes are selected using this interrogation and these are sequenced by means of interpreting a set of sequencing rules. These refer to the mixture of manufacturing processes which have been identified from the interrogation and which are required in order to create the features of the component.

Early versions of this class of CAPP system relied on attempting to automate the variant class of CAPP. This was approached by building fixed sets of sequencing rules into the system. The rules may be extended to cover the capabilities of the machine tools and the tooling available. These can be assessed in order to set operating parameters such as feeds and speeds which will be compatible with say surface finish characteristics such as cusp height.

In order to optimise the feature cutting or forming parameters and sequences, a number of alternative control strategies have been evolved by different CAPP system vendors. These control strategies variously consider process sequences and parameters. These are explored, either to their full finished conclusion, or all possibilities are explored on a broad front one step at a time in order to choose the best next process to evaluate. The generative class of CAPP systems offers a rich application area (domain) for AI software technology and it is likely that more progress will be forthcoming in this important area.

Generative CAPP systems which do not employ AI tend to organise their standard process plan elements using decision tables or decision trees. These are composed of predisposing data conditions, within the model and an associated action to be followed. Decision tables can be used to hold such sets of rules as are required to evolve a process plan. They can be of the form 'IF (condition) THEN (action)' and in this form they are easily edited. However, for other than simple rule sets, they are not easily or quickly capable of being interpreted by the user.

Control engineers have used state diagrams for many years. In this form of rule based control modelling, action states are linked by the main branches, sub-branches and twigs which represent the conditions through which the actions are related. These decision trees are visually clear to the user and can be more easily updated and developed.

Feature establishment

The generative CAPP system must gather information about the manufacturable features and the key parameters which relate to those features from the design data for the component or assembly. This interrogation of the design data must take account of similar features being of different sizes, the physical locations and the relationship to previous and subsequent features to be made. It must also establish the access side of the component from which the feature is to be formed. User interaction is often required to formulate this information within the system.

Machine tool selection

Having established the features which are to be planned, the system must then go on to select appropriate machine tools or other equipment to provide the feature. This selection is based on the information held within the database on the available machine tools and their processing capabilities. Clearly the overall size of the workpiece, the required finish quality and key ratios of the workpiece such as length to diameter will influence the choice of machine tool. The selection will also, in part, depend on the volume run of the component or assembly to be made, the setting up time and the hourly cost for the machine tool. Break even costs can be considered and the choice of machine optimised.

Feature achievement

Having selected the machine tool(s), the next stage within the generative CAPP system is to consider how the feature is to be achieved. The sequence of individual setting up, cutting tool selection, loading operations and machining operations must be determined. Within setting up operations, consideration must be given to the clamping or holding of the part and the setting of machine parameters. These operations will include both metal cutting operations and rapid moves through air in order to position the cutter prior to a cutting operation.

This process involves the determination of the manufacturing strategy using some form of rule modelling. For example, for a turning work centre, all of the turned or bored features which are required to be produced and are both accessed from the same side and are co-axial should be grouped together, in so far as they can be achieved with the same basic setting up operation. Also, cross holes are drilled before principal through holes (to avoid producing burs). Similarly within a milling centre all pocket milling operations from the same direction for a component would be grouped together.

Sequencing of operations must be optimised if an efficient process plan is to result and the quality of the finished component is to be acceptable. Clearly, all operations which can be performed on a particular machine tool need to be grouped together. The remainder of the sequencing will rely on good machining practices as indicated above.

Once the operation sequence has been evolved for each of the set ups, the CAPP system must consider the operating parameters for each of the work centres. Such parameters as the feed rates and spindle speeds will be matched to the material from which the component is to be made, the known state of the machine tool and its tooling. This information will be held within a database concerned with machinability.

Following this process, for each of the groups of features to be produced, and for each of the successive set ups, the generative CAPP system is able to calculate the production time for the component. This time will consist of setting up time, non-cutting movement times and metal removing times. The system is then able to report all of the operations to be performed in a form appropriate to each work centre within the shop floor and beyond.

Process plans must be prepared for each work centre in the optimum sequence to manufacture the component or assembly. A work centre consists of a logical grouping of capital equipment such as machine tools, inspection equipment and the associated labour resources. The capabilities of the work centres are held within a database as part of the CAPP system. Facilities must exist within the system to amend and update the information relating to work centres in order to reflect the development or restructuring as the company evolves.

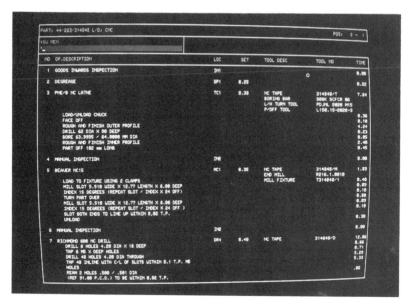

Part of a computer assisted process plan (courtesy of CADCentre Ltd).

Process planning language

Within many CAPP systems, the user employs a special process planning language which is subsequently interpreted by the system. The language is used to encode both the setting up operations and the machining or other operations which are related to any given set up at a work centre. The process planning language allows the user to specify the manufacturing logic and can include some decision making logic using conditional logic statements. By using such language constructions, within a procedural approach, it is possible to generate process plans in a semi-automatic method. These will be in the form of a sequence of work task elements which are represented by standard codes and their associated narrative text. Within these can be embedded variable operational parameters such as length of turning, radius of fillet etc. Using such languages, users can create their own set of manufacturing feature routines. These can be identified by feature codes with associated setting up and operating instructions.

Technical data specific to the company

Within a generative CAPP system there is need to hold a considerable amount of data specific to the conditions within the company. Not least of this data, concerns the technology information which characterises the plant within the workshop and associated tool stores. Such CAPP systems need to be customised with the following types of company-specific technical data. These sets of data are typically held within the following databases within the system:

— Machine capability (speed range, feed range, maximum stroke length, minimum and maximum machinable dimensions, clamping facilities, tolerance capability, tool changing method etc);
— Available tooling (range of cutting tools for each machine tool, dimensions of cutting edge, maximum spindle speed and feed rate etc);
— Data concerning machinability (tables of speeds and feed rates relating to tool types and workpiece materials);
— Data concerned with machine movements which are not associated with cutting or forming metal (times for rapid moves made in the air, dwell times between speed changes, delays for turning coolant on and off, tool indexing times etc).

. CAPP systems make provisions for the user company to build up and edit these various company-specific databases. Such facilities are required to take account of changes in the capital equipment within the various work centres which will vary over time. Similarly, some aspects of the data may need to be changed so that the system can take account of ageing of both capital equipment and tooling.

Numerical control part programming

A special case of CAPP concerns the programming of numerically controlled machine tools (NC machine tools). These machine tools are controlled, in respect of their sequence of operations, by programs of instructions. These programs of machine instructions are referred to as 'part programs'. They are used within the machine tool controller electronics to control the precise operation of the NC machine tool. Thus the part program contains the information to machine, form or assemble a part or assembly. The part programs are either loaded manually into the electronic controller of the machine tool or alternatively they can be down loaded directly into the machine tool controller from a computer which governs the sequence of jobs on the NC machine tools. Programs are sent to each of the numerically controlled machine tools under the control of this computer. This latter technology is known as direct numerical control (DNC).

The use of NC machine tools offers the potential advantage of greater accuracy in the machining process. Each part which is machined within a batch is manufactured to the numerical accuracy determined by the part program and the control resolution of the machine. Thus parts are produced with great consistency and are not subject to the variation normally associated with manual machining.

The capital cost of numerical machine tools is higher than for manually operated equivalents and it is crucial that their capital cost is recovered by achieving high utilisation of such machine tools. In order to meet this objective it is necessary to build up the workload, preferably with high added value components. This, in turn, means that efficient methods of preparing the part programs must be used.

Most numerically controlled machine tools provide some facilities for manually programming the necessary instructions directly into the machine tool controller with the machine tool set off-line. This process involves manually interpreting the manufacturing drawing into the features which have to be machined and then entering appropriate commands which form the part program directly into the machine tool controller electronics, via its associated control panel keys. This program then has to be proved by stepping through the program on the controller and cutting trial pieces, (typically cut from a low cost material such as foam or wood). The proved part program can then be stored on to magnetic media and re-used when required.

The disadvantage of this approach is that the machine tool is normally not available for productive work whilst the new program is being entered and proved out. Clearly, if the machine tool is to be used on long batch runs using relatively few part programs, the overhead of preparing the part programs directly within the machine tool controller is not prohibitive. However, if the mix of production is such that batch runs of the same component are relatively small (in order to keep the NC machine tool busy making parts with added value) many more part programs will need to be created, proved out, stored and rapidly retrieved.

Such cases as this are frequently found in small and medium sized companies including manufacturing jobbing shops. It is often more efficient to prepare the part programs on a separate computer system using a numerical part programming software system. This approach minimises the time taken out from the productive use of the machine tool. It also enables lower batch sizes to be economically handled using numerically controlled technology.

In many companies a number of different types of NC machine tools are used for various manufacturing processes (turning, boring, milling, spark erosion, sheet metal punching and nibbling, bending etc). The demand for part programming, in order to keep the machines busy, can be high. Moreover, as new products pass from the design phase to full production, the load on the part programming function can peak as a number of components which must be planned for manufacture, using numerically controlled machines, need to be programmed. For both of these situations there are benefits to be gained from undertaking the part programming task on a dedicated off-line part programming system. The characteristics of such a system are outlined in the following paragraphs.

Part programming system requirements

Part programming firstly involves extracting subsets of the geometry data held within the design data. Typically, this will be the bounded geometry which, for example, defines the profile of a turned part, or the bounded geometry sets which define the periphery of pockets of a milled component. These sets of bounded geometry will consist of a contiguous sequence of lines, arcs, splined curves, chamfers, fillet radii etc. It is also necessary for the bounded geometry to implicitly or explicitly codify what parts of the geometry are material and what are voids. The part programming system also needs to have a definition of the geometry of the blank material or cast part from which the finished component is to be manufactured. It also must have access to databases containing company-specific technical information on machine types, available tooling, machinability and timing information. These databases were discussed earlier within this chapter. The system should preferably hold graphical information relating to the outline geometry of key parts of the machine tool and tooling.

Most part programming systems are able to accept bounded geometry data from computer aided design and draughting system. The user firstly identifies the sets of bounded geometry which are required to control manufacture as a series of closed shapes, probably set up on nominated layers within the design and draughting model. These are then translated into a form which the part programming system can interpret and are passed directly to the part programming system. Similarly, the bounded geometry of the blank, casting or partly manufactured component can be passed to the part programming system.

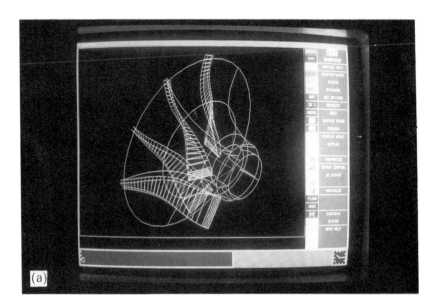

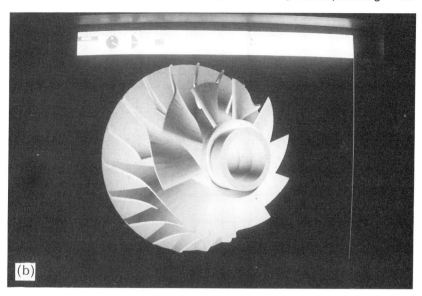

Numerically controlled machining of an impeller blade: a) Defining the surface model;
b) Computer visualisation of the finished impeller blade; c) Foam cut proving model of
the impeller blade and the finished machined component (courtesy of CADCentre
Ltd).

A powerful approach to part programming employs the further use of computer graphics. Such systems are normally structured into a series of application modules. A module is typically concerned with each of the generic types of numerically controlled machining and forming. Typical examples relating to mechanical engineering are as follows:

— Turning and boring;
— Milling, routing and drilling;
— Punching and nibbling;
— Shape nesting;
— Flame, plasma or water jet cutting;
— Bending;
— On-line gauging.

From the system's technical databases, the geometry of key aspects of the selected numerically controlled machine tool can be called up and displayed on the graphics screen. For example if the turning module is invoked, then the chuck, tail stock, tool carousel, tool holder, and bed of the machine tool would be displayed. The part programmer then calls up the geometry of the blank from which the component is to be machined or formed and this is similarly displayed appropriately positioned in the machine tool. The finished geometry should then be superimposed on to the display of the blank. Appropriate tools can be positioned within the tool carousel when their use is invoked by the system user. The associated tool tip geometry needs to be called up from the tool library database. In the case of other machining processes, any clamping fixtures should also be capable of being similarly called up and displayed. This helps to ensure that potential clashes with these fixtures are avoided.

Given these facilities, the part programmer can then select a tool to be used for the start of the machining sequence. The machining operation sequence is then interactively built up together with the rapid and slow moves (in the air away from the workpiece, clamping fixtures and parts of the machine tool). The part programming system should provide a rich set of machining commands, including automatic procedures such as rough cutting of turned profiles and area clearance of pockets. The profile boundaries of areas to be rapidly cleared can be formed by offsetting the geometry of the finished geometry - e.g. turned part profiles, milled pockets, doubly curved surfaces.

Both roughing out machining cycles and finishing cutter operations need to take account of tool tip geometry in order to ensure that the tool is compatible with such workpiece features as fillet radii, slot width and under-cut profiles. In order to confirm this compatibility, the part programming system is required to support a good set of viewing commands including

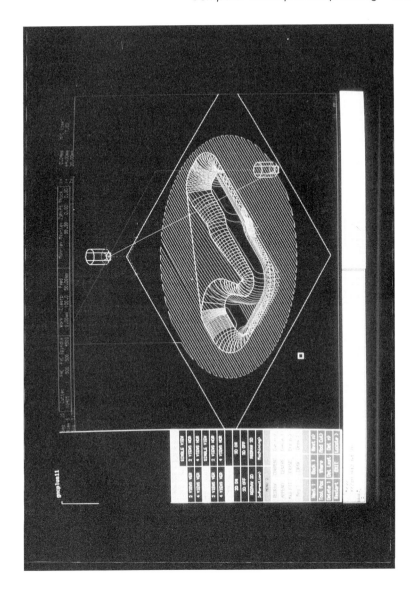

3D simulation of numerically controlled machining of multiple surfaces with gouge avoidance (courtesy of CADCentre Ltd).

zoom and pan. Flexible viewing commands enable the relationship between the tool tip and the workpiece feature to be explored at higher magnification. The system should allow the tool tip to be displayed at any point on the machining or free air movement operations. These provisions allow incompatible tool selection and potential clash conditions to be explored on the screen. They also help the part programmer to make maximum use of the cutting tools and clamping fixtures which are already available within the company. This will tend to reduce the extra costs and delays in having to create new cutting tools and fixtures when existing ones would do the job.

Within sheet metal work numerically controlled manufacture, additional aids are required. The ability to automatically or interactively nest components within standard sheets is of assistance. Nesting aids can help to reduce scrap sheet material and to maximise production throughput. Provision for orientating sheet components with surface patterns is required for such applications as the clothing or furniture industry. Optimising sheet clamping, the punching and nibbling tool selection, and sequence of punching and nibbling or profile cutting operations are again crucial factors in meeting safety, quality and efficiency in production methods for sheet materials.

This approach to part programming allows extensive simulation of the machining sequence away from the machine tool. Three dimensional graphical simulation of all machining operations can be generated on the screen and the part program can be substantially validated prior to final proving by cutting foam, wax or wood parts on the machine tools prior to full production use.

In some companies, more than one generic kind of NC machine tool may be used for making a component. Component manufacture may be carried out at a choice of sites having different makes of NC machine tools or different combinations of electronic controller with similar makes of machine tool. The problem arises of having to have different part programs for the same component. This is because of variations between the control commands which are supported by different combinations of machine tools and their associated numerical control controllers. The problem can be compounded by controllers being able to be configured with different combinations of options selected when the machine tool is installed and commissioned.

Post processors

To avoid these problems most numerical control part programming systems provide for the initial output from the system to take the form of a standardised set of operation commands which dictate the cutter location sequences for a given class of machining or forming. These standardised commands are produced in a computer file for the part often referred to as the cutter location file or CL file. This file can then be translated into precise command sequences for a specific combination of machine tool and each implementation of the associated numerical control controller. The translator for this process is called a post processor and a different post processor is required for each combination. A competent part programming system will support a wide range of such post processors. Some form of semi-automatic post processor generator software module can also be valuable in order to add new post processors to an installed part programming system when a new NC machine tool is installed within the company.

Post processors need to be skilfully prepared if they are to support efficient use of the NC machine tool. Often the part programming vendor, or an associated specialist third party software house, will be responsible for ensuring that the post processors supplied with the part programming system are fully compatible with machine tools and controllers installed within the company.

The use of computer aids, either within the context of conventional process planning, or in association with the use of NC machine tools, can have a major influence on the quality of information and documentation which relate to the detailed production procedures needed to be followed and monitored within the production function of a company. There are clear benefits to be gained from interfacing these computer aids with computer aided design and draughting. However, many companies have derived advantage from concentrating on this area of computer aided engineering as their first investment in technical computing.

9

Computer aided material and production planning and control

The use of computer aids to assist in the planning and control of materials, production and manufacturing processes has been accepted to some degree by many companies. However total implementation of fully integrated computer based systems in this area of the company's work is less well developed. This is particularly true within small and medium sized companies. For many such companies the necessary disciplines implied in such an integrated approach often have not been fully thought through, formally documented and monitored.

Within this chapter, the control of information required to plan and control parts and materials, and the production and manufacturing processes will be discussed. Computer aids are widely available to support the organisation and handling of such information in order that products and systems can be produced in a manner which is optimised to meet the policy objectives determined by the company. These objectives will vary from company to company. The impact which they have on the production process needs to be understood, if computer aids are to be successfully introduced and developed within a company.

Understanding the influence of company priorities

At the outset, a company must seek a consensus within its senior and middle management on the objectives and constraints which apply to the way in which it plans and controls its production and manufacturing processes. In planning these activities, the ranking of the priority of the following company objectives must be discussed and agreed or imposed:

— Minimising stock holding of parts, sub-assemblies and materials;
— Minimising of finished goods stock holding;
— Minimising work in progress;
— Maximising customer satisfaction in respect of delivery times;
— Maximising the use of the capital equipment resources within the production work centres;
— Maximising the utilisation of labour within production work centres;
— Minimising the fluctuation in demand for labour and/or working during unsocial hours or overtime working.

Many company managers, when presented with this list of objectives, will say that they would like to achieve all of them. What joy there would be if this were possible! Clearly, there will be be some mutual incompatibility between the candidate company objectives listed above. For example, customer satisfaction in respect of minimising the delivery time for a product, in part will be associated with the stocking policy for either the finished product or the parts, materials and sub-assemblies from which it is manufactured. Similarly, minimising the stock holding of parts, materials and sub-assemblies is likely to increase the risk of increasing the value of work in progress. This risk will be more severe particularly if there are variations in the lead times for the supply of parts, materials or for sub-contracted work.

Many modern small and medium sized companies aim to meet their production capacity demands with a minimum of overtime working or working in unsocial hours. This can be of importance if there is a high skill content in the work, when the workforce is predominantly of an age where they have strong family pressures not to work unsocial hours. There may also be strong market competition for the required skills. In such cases, the planning of loads on production work centres must be planned with some provision for responding to unforeseen demand and this may impact on the overall optimisation of usage of both skills and capital equipment.

In too many companies often contending priorities are not sufficiently thought through, agreed, documented, communicated or adhered to with sufficient consistency over time. A key advantage of computer based systems in these application areas is the facility to be able to rapidly explore the consequences of changes in operating assumption within the company on a 'what if' basis. By setting the operating conditions within a computer system for production planning and control, the resulting plans and schedules will reflect the constraints imposed by those conditions. It is important to recognise that if these system operating conditions are changed more rapidly than the real company operating environment can respond to, the production function will become unstable and uncontrollable.

Any engineer who has any knowledge of control engineering systems knows that if any system is perturbated more rapidly than its response time, the system will sooner or later become unstable. The problem is that for many small and medium sized companies, this basic fact about control is less understood by many key managers. This is often the case with financial managers and others who often have little scientific training. Such managers may be tempted to rapidly change the operating priorities for the company in response to some factor or other such that the materials and production control systems cannot respond without at least some period of unstable results. If changes in priorities come too thick and fast, any form of control is likely to become impossible.

Say, for example, as a result of interest changes, it is decided to change stocking policy or work centre resourcing. These changes will clearly impact on existing and future customer satisfaction, work in progress values and work centre operational performance. If the changes imposed are large, the production function is likely to become highly unstable for a period until the control systems have settled to the new imposed operating priority assumptions. If the changes imposed are both large and too frequent, production will continue to be unstable and unpredictable. These periods of real operational instability can have effects which cascade up to the company level and may result in loss of market share, lower company profitability, high staff turnover of key staff etc.

Planning and controlling the production function is rather like flying an aircraft; if the control inputs are varied too greatly and too rapidly, the result can only be instability with eventual danger of 'stalls, nosedives and crashes'. It is possible to use computer based materials, production planning and control systems to tune the production function within its operating environment. They can help to forecast difficulties ahead of time, but corrective action must be taken by the managers concerned if danger or disaster is to be avoided.

Within the scope of computer based production planning and control, the system is concerned with modelling information with the following questions:

— What is to be made?
— How many are to be made?

— What is required to make them?
— The sequences within which these things are to be made?
— When are they to be made?
— Where are they to be made?
— What actually happens in the process of making them?

In order to start to plan the production of a product (or a system) using computer based aids, the above information must be held within a number of files or databases. The application software modules interact with this data in order to generate information on parts and materials requirement, work centre loads and schedules, work centre performance, and analysis of product production costs and timescales.

Part numbering

Before discussing the various application modules which are used to handle the planning and control of parts and materials, production and manufacture, the options for uniquely identifying parts and sub-assemblies within a company needs to be considered. Each bought in or manufactured piece part, material, sub-assembly and finished product must be capable of being uniquely identified by a company part number. Information concerning the part number must be held on computer. This is often filed in a part number master file. The underlying objectives within a company's part numbering systems tend to vary from one company to another. Some companies strive to build some form of logical sense into their part numbering system whilst others prefer to enshrine little meaning within the part number.

Part number codes can be entirely numeric or can be a mixture of alphabetic and numeric characters (alpha-numeric). Where an attempt is made to impose some logic to the part numbering system, some part of the part number code may identify the general class of part or sub-assembly (e.g. electrical, mechanical, hydraulic). Similarly part of the code may be used to identify if the part or assembly is made in-house or bought in. Other options can include further refinement of the descriptive class of component (e.g. a type of bearing, bush, shaft, bracket). Yet more refinement to the part numbering system can be added by having part of the code concerned with such part information as a key technical parameter value, a key dimension range, tolerance or type of material from which the part is made.

Such descriptive part numbering systems can, at first sight, appear to be attractive to engineers since they impart some information directly from the structure of the part number code. They can however lead to part number codes which are rather longer than would be needed for other coding strategies. Longer part number codes can result in more frequent errors by those who have to use them. Also, such coding systems need to be controlled and administered in a consistent manner over time if the implied meaning of the codes are to be reliably maintained. There is a danger that different subjective views will occur when new codes are issued by staff, particularly if those staff change over time. The result can be inconsistencies and ambiguities in the part numbering system within the company over a period of years. Such inconsistencies can render the planned interpretation of the part number codes a somewhat hit and miss affair.

For many small and medium sized companies it may be preferable to restrict the degree to which part number codes have any explicit meaning. Regular use of part numbers, even without embodying much technical information within the code, can ensure that staff quickly become familiar with at least those that are regularly used. Perhaps a part number coding system for such companies which merely distinguishes piece parts, sub-assemblies, bought out, manufactured parts and possibly electrical and mechanical components would be optimum.

Range of topics handled within an integrated system

The computer aids for handling information relating to the planning and control of production are wide ranging. The full set of information needed for effective planning and control of both materials and production impinge on other functions within the company. The technology will be discussed under the following topics:

- Parts and material management:
    — Stock control
    — Bills of materials
    — Material requirement planning
    — Works order processing
- Purchase order processing;
- Sales order processing;
- Production management:
    — Production control (scheduling and control of work through work centres)
    — Factory documentation

— Shop floor data collection
— Materials scheduling
— Jobs costing and shop floor performance monitoring

This is an extensive list of topics which must be supported by a competent system of computer aids. The volume of data handled within these activities is large. There is also a high degree of mutual interaction between the application modules required to cover this range. The computer system must support full integration of the functional and data dependencies between these various modules.

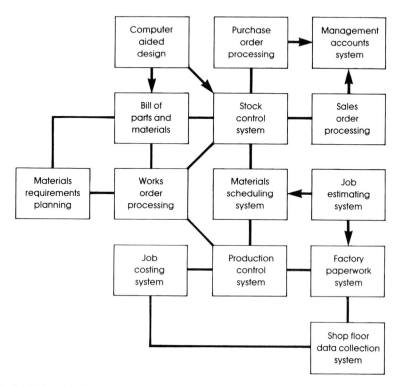

Typical integrated manufacturing planning and control system showing links with external modules.

Parts and materials management

The successful management of parts and materials within a company is crucial. Any system must have the ability to reconcile the company's need to operate within the limits of its working capital and the need to satisfy the needs of its customers at the level of service which has been determined by the collective management of the company. The first of these objectives is in part, related to the management of stock levels of parts, materials, sub-assemblies and finished goods. Both are closely related to controlling material shortages. The area of stock control is central to effective management of parts and materials, assemblies and finished goods.

Stock control

Computer based stock control systems have been widely used for many years. The stock control module provides facilities for the following important activities:

— Up to date reporting of the stock status of all stocked parts, materials, sub-assemblies and finished goods;
— Recording of all types of stock movements in and out of stores areas;
— The provision of aids to support physical stock monitoring, either as routines for perpetual inventory checking or for periodic stock checks;
— The provision of a range of exception reports which can assist management in making decisions on stocking and ordering policy.

The stock control files together should typically support the following data on a part number code:

— Stock group category;
— Part number code;
— ABC classification;
— Item description;
— Unit of measure;
— Stores area location(s);
— Bin number;
— Unit latest replacement cost;
— Date of latest cost information;
— Cost value of stock on hand (based on replacement cost or other criteria);
— Material cost;
— Labour cost;
— Overhead cost components;
— Mean standard unit cost of stock on hand;
— Standard selling price (where appropriate).

— Selling price table for different quantity volumes or order values (where appropriate);
— Free stock balance;
— Allocated stock;
— Stock shortage quantity;
— Stock awaiting inspection;
— On purchase/works order quantity;
— Quantity sold year to date (where appropriate);
— Quantity sold month to date (where appropriate);
— Last stock check date;
— Stock check frequency;
— Minimum stock level quantity;
— Maximum stock level quantity;
— Re-order level;
— Minimum re-order quantity;
— Projected supply lead time;
— Supplier reference;
— Obsolescent stock flag;
— Date of last dispatch of item;
— Special notes.

In addition to the above list of data, flags could be added to identify if the item is a bought out item or a manufactured item and whether the item should be included within the material requirements planning facilities.

If the company operates from more than one location, each of which holds stock, on-line stock enquiries must be supported at each of the appropriate system terminals. Remote access to the other sites' stock file should also be supported, if appropriate. This would allow staff at each of the sites to enquire on stock status at other sites.

Often, a user of the system may not know precisely the part number code for a part or assembly about which they wish to enquire. In this case they may wish to search on strings of characters included in the description field for the part. The use of 'wildcard' characters can be very helpful when enquiring or reporting on items by part number or by description field key words or characters. The wildcard characters can be used with nominated characters to extract information on ranges of parts or assemblies.

Provision should be made within the system, for fields within the stock record which are derived from information provided by stock file maintenance (stock movements), to be automatically adjusted (e.g. mean cost of stock on hand). Notwithstanding this need, as with other highly interactive computer aids, the consistency of response times to all on-line enquiries and particularly to stock enquiries is crucial to the success of the system.

The stock control subsystem should identify and support typically the following types of stock movement transactions:

— Opening stock transaction;
— Placement of purchase orders (on order quantity);
— Stock items inwards from suppliers;
— Stock items inwards from production work centres or sub-contractors;
— Stock item transfers between site locations and between stores areas;
— Stock item issues to work centres;
— Returned stock items from customers;
— Stock items transferred to damaged goods area/supplier;
— Stock items shipped to customers;
— Stock items shipped to sub-contractors;
— Stock item adjustments (following stock checks);
— Stock allocations/de-allocations for sales orders;
— Stock used for internal maintenance;
— Setting obsolescence stock flags;
— Stock losses to scrap;
— Stock item deletions.

The stock movement transactions must be fully integrated with the sales order, works order and purchase order processing modules within the overall system. Of particular importance is the automation of the link between the stock master file and sales order processing. These linkages assist in maintaining accurate and up to date stock status information. Similarly, automatic facilities must be provided for adjusting the various cost price and lead time fields within the stock record file, consequent on the purchase of replacement stock. Using the data extracted from the purchase order processing data assists in updating these costs and lead times.

A number of standard stock reports must be capable of being produced. Certain fields will be obligatory for the reports to have any real meaning and other fields may be considered to be optional and will be selected by the company. Report configuration routines offered by the system should prompt the user to select optional fields to be included (or not) within the report format. Reporting (on-screen or printed) of stock cost information should be under password protection control.

The following stock control reports are typical of those routinely used by many companies:

```
KEWILL SYSTEMS PLC              M I C R O S S              03.03.92

SCS210                    S T O C K   E N Q U I R Y         ver 4.501
                          ===========================


Stock number : PU4432-007              RELIEF VALVE ASSY
                                       MODEL NO 1200
                                       40MM BORE TRIP 50PSI
                                       CASED FOR SHIPPING
---------------------------------------------------------------------------
PRODUCT GROUP CODE ..... V100      ISSUES LAST YEAR .......    343
UNIT OF MEASURE ........ EA         ISSUES THIS YEAR .......    488
BIN NUMBER ............. R5567      ISSUES LAST PERIOD .....     46
                                    ISSUES THIS PERIOD .....     43
STOCK-ON-HAND ..........   149      Max dec.places for costs 2
STOCK-ON-ORDER .........    40      SELLING PRICE ..........     56
ALLOCATED STOCK ........     0      STANDARD COST ..........   34.5
SHORTAGES ..............     1      AVERAGE COST ...........  37.39
FREE PHYSICAL STOCK ....   148      LATEST COST ............  38.25
STOCK-IN-INSPECTION ....     0      LATEST COST DATE ....... 030392
RE-ORDER LEVEL .........    25      MATERIAL COST ..........   15.2
RE-ORDER QUANTITY ......    50      LABOUR COST ............    9.6
LEAD TIME (weeks) ......     5      OVERHEAD COST ..........    9.7
SUPPLIER CODE .......... JASON      VAT CODE ...............
Multiple Locations ..... Y          ABC CODE ............... B
Purchased item ......... N          LAST STOCK CHECK DATE .. 010299
Phantom Assembly ....... N          REGULARITY CODE ........ 4
Bulk Issue item ........ N
Exclude from MRP ....... Y
```

Stock enquiry report (courtesy of Kewill Systems plc).

— Current stock status tabulations;
— Stock movement reports between nominated dates listed by:
    — stock movement type
    — stock category;
— Stock valuation reports;
— Stock status exception reports for a range of part number codes;
— Stock checksheet reports based on stock area code - for use at stock taking times;
— Stock exception reports for items which are in shortage;
    — *Urgent*: stock is over allocated, having negative free stock and for which the 'on order' quantity will not service the required quantity for allocation.
    — *Ordered quantity < Minimum stock level*: the stock is over allocated and the ordered quantity will still leave free stock below minimum stock level.
    — *Order quantity < Re-order level*: the stock is over allocated and the on order quantity will still leave the free stock below re-order quantity.
    — *Below minimum quantity*: this demands a re-order action decision.
    — *Below re-order level*: prompting re-order action considerations;

— Stock exception report for excess stock holdings above maximum stock level;
  — *Free stock > Maximum quantity*: with stock cost.
  — *(Free stock + On order quantity) > Maximum quantity*: prompting cancellation/amendment of the order, with stock costs.
— Stock exception report for slow moving stock (user selectable movement limit);
  — Valuation of total stock for over aged stock for a nominated period;
— Stock exception report for obsolete stock with stock valuation.

As an alternative to controlling stock levels based on operating between minimum and maximum levels, stock items can be monitored by the system based on past demand. The re-order recommendations, when to order, and how many to be ordered can be based on some optimised formula based on historical usage.

Other stock reports should be capable of being generated using a report generator, access to which should be under password protection. The report generator should be easy to use and self documenting. Examples might include price change reports, ABC analysis, variance analysis etc. The precise set of regular reports to be produced by the stock control system will depend on the company. The need to provide acceptable audit trails covering items as they move within the company and beyond will need to be carefully covered.

Bills of materials

Another key element within computer aids for materials management concerns the handling of bills of materials. It is essential to ensure that up to date product structure information is held within the system. The complexity of this data depends, to a degree, on the levels of sub-assembly which are needed to build the company's products and/or systems. The structured bills of materials are also used in conjunction with the cost information on parts and materials to build up cost information for each of the levels of assembly.

The bills of materials data consists of the indented product structure information in the form of part number codes and their associated quantity. Other related data, for example the description of the part, can be extracted from the master information on the part held elsewhere within the integrated system (typically the part number master file).

```
KEWILL SYSTEMS PLC              M I C R O S S  4.501            DATE 03.03.92

BOM210                          PARTS LIST - All Levels           PAGE 1
                                =======================

R2600 1000-000  ROADSTER 1000 26in                  Revision AA ADO/1   26.02.92

     Stock Number   Description                      Quantity   Ref.   Class.
     ------------   -----------                      --------   ----   ------
     FR 1000-000    .FRAME ASSY (R1000)                 1
     FA 1100-000    ..FRAME (26in)                      1                Pur
     HB 1200-000    ..HANDLEBAR ASSEMBLY                1
     HB 1200-100    ...HANDLEBAR                        1                Pur
     HG 1200-200    ...HANDLEBAR GRIP                   2                Pur
     SA 1300-000    ..SEAT ASSEMBLY                     1                Pur
     PA 1400-000    ..PEDAL ASSEMBLY                    2
     PB 1400-100    ...PEDAL BODY                       2                Pur
     FP 1400-200    ...RUBBER FOOT PAD                  2                Pur
     PR 1400-300    ...PEDAL REFLECTOR                  2                Pur
     WA 1000-000    .WHEEL ASSEMBLY (26in)              2
     TA 1100-000    ..TYRE ASSEMBLY                     2
     OT 1100-100    ...OUTER TYRE (26in)                2                Pur
     IT 1100-200    ...INNER TUBE (26in)                2                Pur
     VA 1100-300    ...VALVE ASSEMBLY                   2                Pur
     HA 1200-000    ..HUB ASSEMBLY                      2                Pur
     SP 1300-000    ..SPOKES (13in)                    48                    Ph
     SP 1300-100    ...SPOKE WIRE                      624                Pur
     SP 1300-200    ...SPOKE NUT                        48                Pur
     123            ..WIODET                            2
     AC 1000-000    .ACCESSORIES                        2                    Ph
     AC 1000-100    ..FRONT LIGHT                       2               E.D.C.
     AC 1000-200    ..REAR LIGHT (Red)                  2                Pur
     AC 1000-300    ..REAR REFLECTOR (Red)              2                Pur
     AC 1000-400    ..TOOLKIT                           2                    Ph
     AC 1000-410    ...TOOLKIT CASE (Leather)           2                Pur
     AC 1000-420    ...SPANNER                          2                Pur
     AC 1000-430    ...PUNCTURE REPAIR KIT              2                    Ph
     AC 1000-432    ....CHALK (Powdered)                2                Pur
     AC 1000-433    ....GLUE                            2                Pur
     AC 1000-436    ....PATCHES                        10                Pur
     AC 1000-500    ..SADDLEBAG                         2                Pur
     3162-033       .3 WAY RECP.                        2
     AC 1000-432    .CHALK (Powdered)                   2                Pur

                          End of Report
```

Indented Parts List, Bill of Materials, (courtesy of Kewill Systems plc).

Computer based bills of materials facilities demand rigorous disciplines in preparing the parts lists for assemblies and finished products and systems. These are often indented to show the level of assembly to which the list of parts and materials relate. This indented list allows the costs for product or assembly to be rolled up to compute the latest cost of an existing assembly, product or system. This cost can also be compared with the standard cost. They can also be used to calculate the target cost of a new product, assembly or system.

The bills of materials data entry and amendment facilities should be interactive and controlled under password protected access. In some cases, the structure and content may be similar for a previously defined product or assembly. In this case it is helpful if the system allows previous bills of quantity to be copied into a new structure and to be subsequently amended to meet the requirements of the new assembly. Similarly, such copy and edit facilities are helpful when a product has to be customised.

Another useful facility within a computer based bills of materials module is to be able to add strings of text in addition to the part number code. This information can be subsequently printed on the materials requisitions documents against the part. It can be helpful in identifying such things as component reference information in electrical or electronics assemblies, or piping or instrumentation systems.

User enquiry facilities, from within the bills of materials modules, typically support the reporting of 'where used' information. This assists in identifying the impact of changes due to component obsolescence, shortages or exchanges. In this case, a search of bills of materials structures can be made for occurrences of a nominated part number code. Also, individual parts and material lists, typically can be generated for a nominated assembly or finished goods part number code, and summarised bills of materials (summarised at some level of assembly) produced.

Other standard options include facilities to track and audit engineering changes which are made to the bills of materials, automatic costing at any level of assembly using the standard costs for parts used within the product structure and links to computer aided design and draughting systems. In this latter case, if part numbers are held as 'associated text' within a ballooned assembly drawing model or say an electrical schematic diagram model, automatic data transfer is possible. The parts and materials list for the item represented within the design or drawing computer model can be automatically extracted, and fed directly into the bills of materials module within the material management subsystem.

Material requirement planning

The material requirement planning (MRP) module is an option which can be added to an integrated system when other aspects of control are running well. Its introduction will be possible when the data associated with stocks, purchases and sales orders are known to be reliably held within the system. Its inclusion within a company's computer system should be seen by management as an important goal to aim for. MRP provides more refined control over materials and parts within the company. The MRP module is concerned with determining the need for new purchases or manufacture of items in order to meet the best estimate of demands. With its use it is also able to identify and report existing works and purchase orders which are required to be rescheduled or cancelled.

Within the MRP module, the existing current confirmed sales orders are interrogated, together with planned production orders and any additional forecast demand within the master production schedule (to be discussed later in this chapter). From the information on products and assemblies to be made, held against these orders, and information held within the bills of materials, the total list of assemblies and parts can be exploded. This exploded list can then be scheduled in terms of both purchases of bought out parts, materials and sub-assemblies and schedules of works orders.

This schedule should be treated by the management of the production and purchasing functions, as a list of suggested actions, for their consideration. The actual ordering process, its scheduling and policy strategies are determined by the operational company management. Their decisions can however now be made in the knowledge of the best known information concerning procurement needs. This information has regard to the purchasing and production lead times which have been taken into account by the MRP subsystem.

For example, operational management can elect to use either the net or the gross requirements in their decision making. They can implement ordering actions based on maintaining a minimum stock holding, as held within the stock control system, or some nominated level of demand cover ratio. Alternatively, for slow moving products, or products approaching obsolescence, they may wish to base these decisions on reaching zero stock levels. Similarly, the associated order quantities remain under the control of the operational management. These may be based on multiples of the suggested re-order quantities, held within the stock control subsystem, or they can be based on the projected needed quantity suggested by the MRP system.

Some sales orders, or projected demand levels for certain products, are likely to be more sensitive to being rescheduled. These priority sensitivities can be handled within the MRP system and highlighted accordingly. The parts and materials demands for these priority situations will need to be handled accordingly by the operational management.

Thus, the management of the company retains control over the implementation of decisions which commit financial expenditure on purchases and production runs. They are able to set their own criteria and policies. However, they do so against the background of the analysed need to meet the sales orders, projected demands and any additional forecasts which have been entered into the system.

It was noted when discussing stock control, that provision was made to identify stock items to be omitted from MRP control. Such items as bulk issued materials or low cost fixings and fastening may be excluded and purchased under the controls operated directly within the stock control system. Conversely, non-stocked items, or items not held within a product structure, can be included within the MRP system. These would include such items as factored items or lists of spares to be shipped with the company's products or systems.

The overall objective of using computer based material requirement planning is to model the future stock requirements and work centre loading. This process can ensure that both the overall lead times and stock costs are minimised, by optimising the timing of material procurement actions. If the supporting data is of good accuracy and quality, the reliability of future delivery dates can be improved, customers enjoy a more 'surprise free' future and their probable satisfaction is improved.

By exploring various assumptions on future demands and production loading, the implications for stock holdings, work centre resources and delivery schedules can be optimised within the financial limitation of the company. Much however depends on the ability of the company both to provide good data and to avoid disrupting the actual loading beyond that which has been allowed for in the planning assumption.

Within most material requirement planning modules are normally options to link the MRP module with the works order and purchase order processing modules. Such links allow the automatic preparation of works and purchase order documents in the event that the decision is made to implement the suggestions made by the materials requirements planning module. Similarly, from within the MRP module facilities should be available for producing 'where used' information together with stock status enquiries. Both of these types of enquiry can assist in deciding on which MRP recommendations are best implemented.

Having optimised the works orders and purchase orders which are to be raised, to meet the planned production demand, further modules within an integrated system typically take over the processing of these two classes of procurement documents. The works order processing module will be discussed first.

Works order processing

Works order processing is normally concerned with controlling initiation of production jobs into the work in progress stream. It is also oncerned with calculating the parts and materials need for a particular works order at each stage as the works order passes through the various production work centres. Within a fully integrated system, various options may be taken as to which of the range of modules are implemented by a company. Most companies try to phase the introduction of the various system modules over a period of time. It was seen earlier that MRP is often implemented at the later stages of computerising the production function.

Hence, in the case of works order processing, the functions performed will often depend on which other modules have been implemented within the company. Works order processing may be implemented in some typical systems in association with the stock control and bills of materials modules. If this combination of modules is implemented, the facilities outlined in the following paragraphs should be sought from the system.

For this combination of modules, the main facilities provided by the works order processing module are likely to be concerned with handling the materials need of the works order, and tracking the receipt of the work resulting from the works order back into stores. The works order processing module in this case should provide interfaces with the stock control and bills of materials modules of the system. This allows the works order module to calculate the parts and materials requirement and to forward allocate them on the stock control system. This forward allocation action identifies the need for stocked items in order to satisfy the works order. Earmarking stock items in this way reduces the quantity of free stock which in turn, may trigger purchasing or other ordering action within the stock control system.

```
KEWILL SYSTEMS PLC            M I C R O S S - 4.15              DATE : 12.03.92

Report WOP330                 P I C K I N G   L I S T S         PAGE :      1
                              ==========================

                              Picking List number : 231

                                                      WORKS      QTY TO    TOTAL TO
    LOCATION   STOCK NUMBER   DESCRIPTION        UOM  ORDER NO  BE ISSUED  BE ISSUED
    --------------------------------------------------------------------------------
               3162-033       3 WAY RECP.         EA  W6552        200        200
               123            WIODET              EA  WQ0123        27         27

    AA1010     WA 1000-000    WHEEL ASSEMBLY (26in) EA W6552       200        200

    AA2015     FR 1000-000    FRAME ASSY (R1000)  EA  W6552        100        100

    AB1998     AC 1000-410    TOOLKIT CASE (Leather) EA W27361      37         37

    AC1087     AC 1000-420    SPANNER             EA  W27361        37         37

    AC1654     AC 1000-432    CHALK (Powdered)    EA  W6552        200        200

    AC2041     AC 1000-400    TOOLKIT             EA  W6552        200        200

    AD1018     AC 1000-500    SADDLEBAG           EA  W6552        200        200

    BA1066     AC 1000-430    PUNCTURE REPAIR KIT EA  W27361        37         37

    BA4069     AC 1000-200    REAR LIGHT (Red)    EA  W6552        200        200

    BB5032     AC 1000-300    REAR REFLECTOR (Red) EA W6552        200        200

    CA2025     SP 1300-000    SPOKES (13in)       EA  WQ0123       648        648
```

Part of a computer generated picking list (courtesy of Kewill Systems plc).

Prior to issuing the works order to the shop floor, a trial kitting list can be produced. This can be used in order that, not only stocked items but also any non-stocked or special items can be checked for availability. In the process of interacting with the stock control system, the works order processing module will identify shortages of parts and materials and these can be reported by exception at an early stage. Remedial action can then be initiated in good time. Similarly, non-stocked items can be checked for availability within the kitting area. When the parts and materials needed for the works order can be satisfied, and shortages have been cleared, the works order can be issued to the shop floor. The parts and materials requisitions for issuing blocks of material to service the works order can be automatically generated together with picking lists. The picking lists group stock items within their stores locations for different works orders. Hence stores movements to work centres for works orders can be serviced efficiently.

The parts and materials requisitions thus generated, as part of the shop floor documentation, should be cross referenced both to any picking list produced, and to the works order. It should typically identify the following information:

— Administration information:
  — Date printed
  — Requisition reference number
  — Works order reference number
  — Due date for completion of the works order
  — Due date for issuing the materials
  — Part number of the highest level of assembly to which the works order refers
  — Description of the highest level of assembly
  — Quantity of this assembly ordered on the works order
  — Quantity of this assembly outstanding
  — Quantity of this assembly already issued
  — Quantity of this assembly to be issued
  — Work centre/sub-contract supplier reference code to which the parts, assemblies and materials are to be issued
  — Cross reference to picking list document number
  — Special notes (e.g. avoid magnetic fields, keep upright);
— For each part, sub-assembly or material type:
  — Item reference (for this document)
  — Part number code
  — Part or material description
  — Unit of measure (for issuing)
  — Stores bin location reference

— Quantity to be issued (multiple of unit of measure)
— Space for entering the actual quantity issued
— Notes for special instructions (e.g. check protective covering).

The picking list is used to collate items to be withdrawn from stores in order to service a number of works orders. The sequence of the list should relate to the sequence of physical bin locations within the stores area and will typically identify the following information:

— Location bin number code;
— Part or material code number;
— Part or material description;
— Unit of measure for issuing;
— List of works orders and associated quantities;
— Total quantity to be withdrawn from stores bin location;
— Work centre/sub-contract supplier reference to which items are to be issued.

The works order processing module provides for receiving the finished work from the shop floor into stores. It also supports enquiries into the status of individual works orders, stock status of parts and assemblies, and parts lists. The module should further provide a summary report of all works orders which have been active between selected dates or which are of a nominated status type.

If the modules needed to support full production control, job costing and full factory paperwork (including process planning) are added to a company's computer aided production system further functionality can be supported within the works order processing module. In particular, when a works order is entered, the associated information concerning the manufacturing routing (the sequencing of the job through work centres), can be loaded into work in progress and the cost of the labour content of the job, based on the planned times, can be calculated. In this case, the works order is queued into the works order processing module. This enables a works order to be selected by the operational staff, for the printing of job route cards and the operation tickets for each work centre within the routing. Also, with these additional modules, further reporting options become available. These would include enquiries on work centre labour rates and further summary reports on the works order, including costing details.

Purchase order processing

The use of computer aids to process purchase orders is an area where the take up has been patchy. Much depends on the volumes of purchase orders handled within the company and the complexity of the supply situation. The objective of such a module is to assist in the rapid production of purchase orders and their subsequent chasing, reconciliations with associated documents and progressing the order through to final payment of the suppliers invoice.

The purchase order processing module needs to interface with the stock control subsystem and also with the company accounts system. Flexible facilities should be sought in respect of recovering full details on stock items and their preferred and alternative suppliers. Search facilities using 'wild card characters' (as outlined earlier) are often helpful in identifying parts for which there may be some possible source of ambiguity or with which the user is less familiar. The purchase order processing module must also be capable of handling non-stocked items. For some companies, it may be important to handle purchases supplied direct to another site rather than into stores - e.g. for situations where final assembly, erection and commissioning has to be performed on a customer site (e.g. conveyer systems, cabling and piping).

Full integration with the stock control module is needed in order that pricing information is updated and supply status is passed to the stock system. A further need is to ensure that conversion, between the standard unit of measure which is used within the company (for the purpose of issuing), and that which is used in the purchase of the item from the supplier is adequately handled (e.g. single items and boxes of items).

The purchase order processing module should typically support a comprehensive set of integrated functions associated with raising and processing purchase orders. In summary, the module should normally provide facilities for the following basic functions and information associated with the purchasing process:

— Master supplier information;
— Purchase order status file;
— Purchase invoice and credit note processing;
— Links to the purchase ledger within the accounting functions;
— Generation of goods received notes;
— Goods returned documentation;
— Purchase order adjustments;
— Remittance advice printing;
— Aged creditor analysis;
— Cash flow projection for supplier payments;

— Schedule of open purchase invoices;
— Payment release for purchase invoices with control parameters;
— Payment printing;
— Overdue delivery chase lists.

Supplier information

The supplier information typically should contain at least the following data fields:

— Supplier reference code;
— Supplier company name;
— Supplier address for orders;
— Supplier address for remittance;
— Supplier telephone number(s) for orders;
— Supplier phone number for remittances;
— Supplier fax numbers;
— Terms of business agreed with the supplier:
    — Discount rate
    — Discount days
    — Net days
    — Method of payment (cheque, bank transfer etc);
— Supplier priority rating;
— Business volumes:
    — Credit limit
    — Current balance
    — Purchase volume month to date
    — Purchase volume year to date;
— Date when first registered with the supplier;
— Date when last ordered from the supplier.

Audit trail facility

The purchase order processing system should provide a full audit trail facility to enable all purchase invoices and credit notes to be traced and audited. Typically the following information on purchase orders should be accommodated within the module:

— Purchase order reference;

- Issue number;
- Originator's initials;
- Original order date;
- Amendment authority initials;
- Amendment date(s);
- Supplier reference code;
- Supplier name;
- Supplier order address;
- Supplier contact name;
- Supplier telephone number;
- Target delivery date;
- For each item on the purchase order:
    - Item number
    - Company part number code
    - Part or material description
    - Item order quantity
    - Expected unit net price
    - Actual unit net price
    - Received quantity of each item with date received
    - Rejected quantity of each item with date returned
    - Item status (overdue/delivered/part delivered);
- Total net invoice value payable;
- Total taxes payable;
- Purchase order status (pending/overdue/cleared for payment/paid);
- Date paid.

If the company needs to handle direct supplies to customer sites the following additional information should be handled by the module:

- Direct supply/to stock flag;
- Customer sales order number reference;
- Customer name;
- Direct supply delivery address.

The purchase order processing module should support routine reviews of purchase orders, sufficient to promote efficient chasing of overdue purchase orders, and rapid access to the status of orders and deliveries of items contained on them.

Provision must be made to clear down purchase orders in the event that the user elects to do so. This will be necessary if full delivery of all items is not finally received in their full order quantity shown on the original order. In such an event the 'quantity on order' for the part number code must be adjusted accordingly within the stock control system.

Typical documents produced by purchase order processing

The module needs to provide the documents and reports listed below. As with the other report generating modules, the user should have facilities to simply configure the format of reports. Optional fields, may need to be added to the normal essential administrative fields within the reports. The types of documents and reports which are typically produced by this module are as follows:

— New purchase orders (uniquely numbered - with the originating site identified);
— Amendments to purchase orders;
— Supplier listings using a selection of optional selection criteria (e.g. short form reports containing supplier reference codes, addresses and phone numbers);
— Schedules of active purchase orders, with status data on items and completed order;
— Goods received notes;
— Rejection notes;
— Concession notes;
— Listing of purchase invoices cleared for payment - with values;
— Purchase invoice payment cheques with remittance advice notes;
— Detailed aged creditor analysis reports - with the account position over say the last four periods;
— Purchase orders placed - with forward cash projection;
— Analysis of supplier performance.

Sales order processing

A similar strategy to that which applies to purchase order processing is adopted in the area of sales order processing. The use of computer aids in this area of the company's operations depends, to a degree, on the pattern of sales within the company. The attraction of using computer aids to assist in the management of information relating to sales is stronger where there is a high volume of sales transactions. Many small to medium sized companies have implemented computer based sales order processing - often in isolation from the production and manufacturing activities.

To gain maximum benefit, the sales order processing module needs to be intimately integrated with both the stock control and company accounting functions. It provides the point at which the commercial demand for the company's products, systems and services can be fed into the production and manufacturing systems. The module allows the automatic generation of the documents and reports directly associated with the selling activity of the company. In particular, it should provide for the following output to be produced having input the details of a sales order from a customer:

— Sales order acknowledgements;
— Shipping and advice note documentation;
— Sales invoices;
— Sales credit notes;
— Collation of different customer account numbers (for different delivery addresses) to a common customer accounting address - for the preparation and sending of period end statements;
— Receipts of customer payments from their accounting headquarters site;
— Listings of active sales orders, sorted by user selected keys, including forward orders and their allocation status;
— Short form lists of customer delivery address and contact data within customer invoicing references;
— Sales analysis reports.

The sales order processing module should provide for holding and keeping up to date administrative information about the company's customers. The nature of this information is very similar in style to that identified earlier in connection with suppliers, within the purchase order processing system. It is concerned with supporting good communication with the customer, recording the terms with which the company currently does business with the customer (including discounting arrangements), the current credit limit of the customer, providing cross references to the sales ledgers within the accounting function and the recording of summary sales history information for the customer.

Customer credit

During the entry of sales order information for a customer their credit limit should normally be monitored. If, after each detail order line entry, the updated current credit amount exceeds the customer's credit limit this should be clearly indicated to the user. Only under password control should this or further sales items be accepted on to the system for the customer. The customer should normally be placed on 'temporary stop' at this level of credit exposure. Further action, under password protection, should be taken by the company's credit controller, within the accounting function, if it is thought to be prudent to place the customer on a 'permanent stop'. The customer would then remain on 'stop' until the credit controller confirms that the customer's payments had brought outstanding credit below his or her credit limit.

Within the accounting system, for a customer on permanent stop, the customer account must remain open in order to accept cash payment from them. It is also helpful if the associated accounting system modules support forward posting of sales invoices, credit notes and remittance receipts to appropriate ledgers at period end. Such comprehensive forward posting facilities are often needed if the credit control system is to work satisfactorily and to avoid peak operational loads after closing ledgers at period ends.

Rapid access should be available from any of the on-line screens within the sales order processing to both the data relating to customer administrative information and to any particular sales order record for the customer. Often a customer, when making a telephone enquiry about their order, cannot quote their customer reference code. Alternative entry to their records must be provided. Searching should be supported on some combination of the customer's post code, telephone number, address or customer name. As with other searches, it is also helpful for the module to support wild card characters within such searches. It is essential that on-screen access to any entry within the customer's information, or the sales order details should be rapid.

Sales orders invoices and credit notes should be capable of being tracked and collated within the module and transferred to the sales ledger and reconciled with the day book entries within the accounting system.

For some companies it may be necessary for the sales order processing module to support forward orders with facilities to call off goods over time. This need would benefit from being integrated with the material management modules of the overall system. Another frequently required facility is to be able to process pro forma and 'charge only' invoices. Pro forma invoices are normally used when the customer has not yet been assessed for a credit rating or has possibly had a history of poor payment under previously existing credit terms. Charge only invoices need to be handled where the company provides services without the supply of goods, for example the provision of services such as training or planned maintenance. It is important to ensure that such sales transactions are covered by the application software system.

On-screen data

On-screen enquiries of any individual sales order must be rapid - sufficient to support telephone enquiries from customers. The on-screen data relating to any given sales order should typically provide the following data:

— Customer invoicing reference number;
— Customer delivery site reference number;
— Customer name;
— Company sales order reference number;
— Customer related discount table for different groups of products;
— Customer purchase order reference number;
— Customer delivery contact name;
— Sales order date received;
— Scheduled delivery date (with provision for forward orders);
— Customer delivery address;
— Customer delivery telephone number;
— Customer sales value for calendar year to date;
— Customer credit limit;
— Customer on 'stop' indication;
— For each item on the sales order:
    — Item number
    — Company part or material number code (where appropriate)
    — Unit of measure for sales
    — Quantity ordered (multiple of the sales unit of measure)
    — Quantity dispatched (multiples of the sales unit of measure)
    — Unit selling net price appropriate to customer
    — Sales tax charged on item
    — Date when item dispatched
    — Delivery note reference number
    — Anticipated date for dispatch (if not yet delivered or forward ordered);
— Delivery charge;
— Carrier reference code;
— Total invoice value excluding sales taxes payable;
— Sales taxes (e.g. value added tax) payable;
— Sales order status (pending delivery/part delivered/supply complete/part paid/fully paid).

Sales invoicing procedures

The sales invoicing procedures should take account of the discount structures which apply to an order for the particular customer. Sales discounts will typically use the data held within both the sales order processing module customer data, and the part number specific discount structures held within the stock control system. If a credit customer enjoys discounts, as determined in their customer record, these should normally take precedence over quantity break or product group value discounts held within the stock control data. Exceptions may be needed for stock item special prices which may be listed in the customer record.

Provision should be made for entering sales order amendments and cancellations. These entries should automatically update the stock control data by transferring quantities between free stock and allocated stock and vice versa where appropriate. It is also useful to add textual notes to invoice or dispatch documents at the time when the document is raised. These notes may be specific to the customer (e.g. special delivery instructions) or they may be general announcements such as notification of annual shut down dates for holidays, special promotions, etc.

Remittance processing should be capable of referring to multiple sales invoices. Partly paid customer accounts/invoices should be clearly indicated. Typically, invoices should be cleared in historical sequence. Forward posting of remittance receipts at period ends is also desirable in order to facilitate the correct operation of the credit control system.

Sales order data

Editing of sales order data must be rapid. It is helpful if there is a facility to cancel nominated sales orders (under password protection) by 'a quick kill' facility, rather than having to issue a series of credit notes. This would only apply if no deliveries had been made against the sales order, prior to its cancellation.

For some companies, a facility may be required to validate that any goods returned by customers had been recently supplied to them by the company. Thus, it may be required that a screen report can be rapidly generated, based on a combination of customer name or customer account number and company part number code. The purpose would be to display the details of supply of the nominated parts and to confirm their supply, with the last supply date for the item to that customer.

Comprehensive sales order analysis facilities are normally required and provided. Such analysis data provides management with key data for assessing sales trends, sales performance and monitoring the profitability of business both in respect of customers and items or groups of stock items sold. The company needs to assess which combination of sales related parameters will be most helpful in their particular case. The following parameters can be helpful in different combination for sales analysis reporting:

— Sales representative code;
— Customer delivery site reference code;
— Customer invoicing reference code;
— Invoice sales value (excluding sales taxes);
— Stock group category code;
— Dispatch date.

It is to be preferred (under password control) that such analysis keys are combined using the standard facilities of the sales order processing and analysis module, without recourse to using programming techniques. A standard requirement is to produce definitive sales profitability analysis reports. These are often collated by customer invoicing reference code or customer delivery reference. Net sales value and profitability by stock group code, within monthly, quarterly, and annual time frames are typically reported. Such reports are helpful in assisting a company in monitoring pricing and discounting policy. An equivalent profitability report, collated by product part number or stock group codes across all customers, or all customers within a market sector or territory can be useful.

These profitability reports can provide powerful monitoring of product profitability across the company's market sectors. The reports should be in summary form without the need to list all sales order detail lines. The module should highlight loss making business within these analysis reports. Facilities should also be considered for generating graphs, pie charts and histograms from sales analysis data. This can be done by extracting the data in standard ASCII character form and passing it to a conventional business graphics software package.

Production management

The efficient management of the production processes within a company is an area within which computer aids have enjoyed great success. Good management and control within the production function can lead to shortening of production lead times and optimisation of production resources to meet the policy objectives of the company. In order to provide the environment for comprehensive production management, the following topics need to be considered.

Each of these topics, within the ambit of production management, is typically handled by individual software modules which themselves are linked to form an integrated production management system.

Production control

The production control module is often central to such an integrated system. It is concerned with allowing the user to schedule and monitor jobs passing through the company. The schedule may also allow the user to specify the amount of resources which are to be made available over the scheduling period within each of the work centres.

A competent computer aided production control module will enable the user to monitor all the jobs which form work in progress within the company. Enquiries to this module will provide up to date information on which jobs are partially completed together with the outstanding operations which need to be completed within the work centre(s). Similarly, completed jobs and those which are overdue can also be monitored and reported.

The user typically describes the resources which are available within each of the work centres. These resources are expressed in terms of the number of manufacturing units (e.g. machines of similar type), together with the number of hours which are available within which the work can be handled. The available resources can be specified say week by week within a planning period and within each of the work centres.

Jobs which are required to be loaded will be of three general types. Firstly, there are those which are repeat jobs for which the loads and work centre routing is already known, from previous directly similar jobs. In this case the loads and sequence of routing can be extracted from library held data and loaded on to the work centre schedules. Similarly, if the job is a variant of a previous job, the nearest equivalent job held in the library can be loaded on to the work centre schedules and the user can edit these loads in order to take account of the variations. Finally, the job may be one which is not similar to a previous job and which is not likely to be repeated in the future. In this case, the loads and routing information throughout the work centres can by entered directly on to the work in progress schedule.

In the scheduling process, the user must decide whether the planning and scheduling is to take account of the 'finite capacity' which it is planned to have available. This finite capacity is expressed within each of the work centres over the planning period held within a 'factory calendar'. Alternatively, the jobs may be planned on the basis of so-called 'infinite capacity'. In this case the module will indicate the amount of resources which must be made available within the work centres in order to meet the required delivery date for the jobs which are loaded on to the work in progress schedule.

If finite capacity planning and scheduling is used, the priorities of each job to be loaded must firstly be determined by the system, based on its required delivery date. This then determines the sequence for scheduling the jobs through each of the work centres. Any resulting incompatibilities between delivery targets and the available resources can thus be identified. Remedial action can be taken by increasing the planned resources or by rescheduling jobs in order to minimise late deliveries. These alternative production plans can then be explored. This 'what if' type of exploration typically will need to take account of unplanned loads within some work centres. The provisions for handling unplanned loads are in order to cover emergency work or possible rework. Some assumptions must be made as to what sensible load should be imposed by such activities for each work centre.

When the production master schedule has been optimised, a number of reports and documents can be automatically generated from the production control module. The data for the shop floor documentation, needed to meet the target production plan, can be passed to the shop floor documentation module. As well as an overall production plan for the planning period, 'work to lists' can be produced for each work centre in order to instruct the work centre as to its planned work program for the period of the plan. Also, work loading reports can be generated for each work centre. These identify how the work load for manufacturing units and manpower vary over the planning period. Such reports will automatically identify any overload or spare capacity conditions.

If overloads are identified these become known before the work is due to be carried out and remedial actions can be taken. This can be achieved by planning for overtime working, sub-contracting work out or rescheduling jobs to periods of lower loads - with possible consequences on overall delivery time. Within some systems, there are also facilities for outputting the resulting data in a PERT type network presentation. This can show the dependencies of work centres within the mix of work comprising the work in progress.

Job status information is fed back to the production control module from the work centres as the work program is executed. The feedback of job status information is used within the system to record and monitor the progress of the implementation of the production plan. This progress data is either entered into the system from completed documents, passed back to the

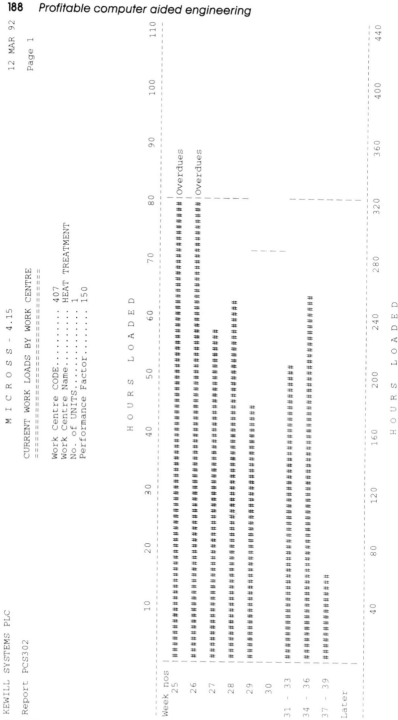

Current work loads of work centres (courtesy of Kewill Systems plc).

production control function, or alternatively captured rapidly from data entered directly on-line to the system via shop floor data collection terminals, sited at the work centres.

From within the production control module enquiries typically can be made concerning the following types of information:

— The overall production plan within the factory calendar;
— The current data held on each of the work centres in terms of the production units and the labour resources available;
— Modifications which may have been made to the work centre facilities;
— Summaries of the jobs which are planned to form the work load and work in progress throughout the planning period;
— Current work in progress within the work centres;
— Libraries of load and routing information on previous jobs.

Normally, reports can be demanded on the status of any nominated job, or any overdue jobs. These are in addition to the work centre 'work to' lists, the list of job priorities and work centre loads. They may be extended to cover estimates of delivery time, based on 'what if' conditions. In the case of using infinite capacity as the basis for production scheduling, the capacity requirements which must be provided in order to meet the required delivery schedule can be ascertained.

Factory documentation

Once a production plan has been finalised for a given planning period, all of the factory documentation needed for guidance of the production work centres, in order to implement the production plan, can be produced. This is often performed automatically from a separate module concerned with shop floor documentation. This module typically produces the following shop floor paperwork:

— Job routing card;
— Trial kitting and kitting lists;
— Job parts and materials requisitions;
— Work centre operations cards;
— Job final test and inspection card.

These documents may either take a standard format, determined by the preprinted stationery for the proprietary software system, or they may be customised to meet the format and layout needs specified by the company.

Sample company documents produced by an integrated production planning and control system (courtesy of Kewill Systems plc).

Often it is helpful to include standard paragraphs of text to highlight particular instructions within these documents such as safety instructions or inspection checks. Typically these can be built up within the factory documentation module and retrieved by specifying associated nominated key word or codes. Similarly, there are provisions made within the module to identify associated drawing number references for manufactured components or sub-assemblies.

When the factory documents are needed to be printed, in order to issue the job to the appropriate work centres, the information concerning the quantities, work centres, planned times and scheduled due dates are extracted from the information on the job and from the production plan. This data will have been produced by the production control module and, optionally, from the process planning system previously discussed.

Where the company has implemented a bar code system within a shop floor data collection system, these can also be added to the work centre job tickets. Bar codes can facilitate efficient and error free feedback of progress information back to the production control module. Alternatively there are fields provided on the job cards for the work centre supervisor to enter the actual start and finish dates and actual setting up and operation times. This data is in turn passed back to production control and entered into the production control module to update the job status information.

Trial kitting lists

The trial kitting list allows parts and materials to be checked against the requirements of a particular job. Potential shortages can thus be identified and remedial action taken, where possible, ahead of the time when parts and materials are required to be issued to the work centres for the job. This facility can be extremely valuable in ensuring that important sales orders are produced on time. Kitting also allows parts and materials to be earmarked for a particular job. Any shortages for a job can be recalled from an associated shortage list and subsequently issued (from the shortage list) to the job as soon as the parts or materials become available within the stores.

Parts and materials requisitions

The parts and materials requisitions for each of the work centres for the job are generated from the bills of materials for the assemblies implied in the production run. The required quantities are extended for the quantity of product or assembly which is to be manufactured. The required issue dates for the parts and materials are included within the materials requisition list. Similarly, if bar codes are used within the company's stores, these can be generated on the parts and materials requisitions.

The use of such a module ensures that data issued to the work centre is compatible with the latest information regarding the assembly or product to be manufactured. It is also presented in a clear, legibly printed, consistent format and standard manner, to the benefit of both the stores and the work centres. These characteristics are particularly important when implementing quality standards within a company, in order to meet appropriate national standards (e.g. within the UK BS5750).

Shop floor data collection

The siting of shop floor data collection terminals can be related directly to individual work centres or a terminal can serve more than one work centre. The data fed back by work centre supervisors provides progress information directly into the production control module. It allows work centres to directly book on and off jobs identified on their 'work to' list. This direct access provides more rapid feedback of information than having to process returned job tickets, upon which work centre actual performance data has been entered. If the factory documentation and the shop floor data collection system both support bar codes, the risk of data entry errors can be reduced. This is because much job identification data is already pre-coded in bar code form on the shop floor documentation at its origination.

Material scheduling

Material scheduling can be of value within companies which tend to manufacture from raw materials in order to make components or products, rather than assemblies or finished higher level products. For such component manufacturers there is little content within the conventional bill of materials structures. A materials scheduling module within a computer system allows a schedule of both bulk materials, and possibly associated manufacturing tooling, to be incorporated within the production schedule work in progress. The module typically allows the total materials to be computed for a production batch run. This material requirement is then compared with the current stock holding for the material and the appropriate quantities of material can be allocated against the production demand.

This facility allows forward warning of bulk materials requirements in order to service the planned production run and the future planned usage is recorded. The information can assist the buying function by providing early warning of needs. Such advanced warning may assist the buyer in optimising the supply of bulk materials and allow him to negotiate improved terms of supply, perhaps by forward ordering the materials with a planned call off from the supplier.

Job costing and shop floor performance monitoring

Having provided all of the production forward planning and monitoring mechanisms within an integrated computer system, at both the job level and the work centre level, it is a natural step to extend this information to provide extended costing information. Job costing and shop floor performance monitoring is typically the function of yet another optional software module within a comprehensive production system.

This module provides for rigorous recording of actual set up and operation times, sub-contract costs and material costs as the job progresses through the work centres. It also provides for collating the costs assigned to a job within the production process. Some activities within a work centre cannot be legitimately assigned to a job, for example cleaning down, routine monitoring of performance parameters, machine calibration routines or purging a machine. These costs are indirect costs. They should be considered as part of the operating costs of a work centre, to be recovered within the overheads of the company in some manner. They will however have an effect on the overall operational financial performance of the work centres within the production function.

Costs typically can be be booked directly into the job costing module for the job, or they may be transferred from other modules within an integrated system. Such transfers of costs to the job costing module would arise from the following types of transactions:

— The processing of job related materials requisitions through the stock control module;
— Parts and materials returns from work centres;
— Parts and materials scrap dockets;
— The feedback of actual set up and operation times from work centres;
   — From on-line shop floor data collection bookings to the job
   — From batch processing of job tickets returned to production control from the work centres;
— Material purchases shipped direct to customer sites;
— Sub-contracted works and services;
— Other job expenses.

This data is constantly updated as the job progresses through work in progress until the finished goods are either shipped to the customer or booked into finished goods stores. Thus, this module is inserted in the link between shop floor data collection or batch processing of completed job tickets, and the production control module. It forwards the information on, to update the status of the job within the work in progress model maintained by the production control module.

In this manner, the actual costs of labour, materials, machine time and other costs associated with the job, are assembled within the job costing and shop floor performance monitoring module. It also has access to the planned or standard material, labour and other costs for the job. The module consequently can report variances from the standard costs for the job at any time, as it passes through work in progress. It can thus alert the production management to discrepancies from the planned or expected costs of the job. In addition, since the module accumulates the added value of the jobs as they

```
KEWILL SYSTEMS PLC                              M I C R O S S - 4.402                        12 MAR 92

Report JCS300                                  J O B   C O S T   S H E E T
Page    1
                                               ================================


JOB NUMBER : 22725              DESCRIPTION : C60m/8m CUTAWAY COULTER   CUSTOMER : D.KURSTJENS    STATUS : In Progress
------------------------------------------------------------------------------------------------------------------
                                                                                      Planned     Actual
Job Category Code       : 100       Planned Total Hours  : 221.8       Labour Cost    :  1428.86   1936.71
Customer A/C Number     : DK0001    Planned Hours To-date:  17.1       Material Cost  :   750.00    448.00
                                    Actual Hours To-date :  51.6       Expenses       :   250.00    150.00
Date Required   (wwdyy) 50.5.80                                        Overhead Cost  :   728.66    225.81
Date Completed  (wwdyy)             Batch Quantity       :    80
Date Despatched (wwdyy)             Quantity Completed   :     0        Total Batch    :  3157.52   2760.52
                                    Quantity Invoiced    :     0        Total Unit     :    39.47
Overhead - Labour       : 30%       Scrap Allowance      :    5%
Overhead - Material     : 30%                                          Quoted Price   :    39.47
Overhead - Expenses     : 30%       Invoiced Value       :  0.00       Invoiced Price :     0.00


* * *  L A B O U R  * * *

Netwk ref  Drawing number                    Quantity  Plan hrs   Plan cost     Status
========================================================================================
  0-0       153318       760m/8m CUTAWAY COUL    80      221.8     1428.86   IN PROGRESS

           OP  WORK  EMP  PAY   QTY   QTY  ----------PLANNED-----------  ----------ACTUAL-----------
           NO  CNTR  NO.  RATE  PROD  SCRAP SET-UP  RUN   COST    SFD   SET-UP  RUN   COST   DATE
VARIANCE

            1  1000  100  1.00   80          0.25   4.00  25.85 42.3.80  0.25  4.20  27.06  42.3.80 C
1.21-
            2  1002  101  1.00    0          0.50         3.23           1.00        6.45  42.4.80
3.22-
            2  1002  100  1.00   40                2.00  12.04                 2.10  12.64  42.4.80
0.60-
            2  1002  100  1.00   40                2.00  12.04 42.5.80          2.15  12.94  42.5.80 C
0.90-
            3  1001  102  1.00   80          0.33  4.00  29.43 43.1.80   0.35  4.00  29.56  43.1.80 C
0.13-
            4  1004  102  1.00   75      5    0.25  3.75  25.65 44.1.80   0.20  3.80  25.61  43.5.80 C
0.04
           10  1000  102  1.00   80                      0.00            1.00  2.50  21.74  25.1.82 C
21.74-
           20  1000  100  1.00   80                      0.00            1.00  2.00  18.70  25.1.82 C
18.70-

                   TOTAL:        80      5    1.33 15.75 108.24          3.80 20.75 154.70
46.46-


* * *  M A T E R I A L  * * *
                                                                             Date
Netwk ref  Stock number                    UoM   Qty    Unit cost    Value  dd.mm.yy
====================================================================================
           CWC-1000   STEEL SHEET           EA    80      5.600      448.00  25.07.80

                      TOTAL:                       80                448.00
```

Job cost sheet (courtesy of Kewill Systems plc).

evolve, and it knows what jobs remain as work in progress, it can, at any time, provide an accurate estimate of the true value of this work in progress.

At the conclusion of the job, a full analysis of job costs can be reported, together with variances against the standard or estimated cost for the job. These job costs can be broken down by each work centre's labour, machine time, material costs and other cost headings. Such job costing information can be an extremely powerful source of data for initial estimating of similar production jobs in the future.

Just as cost data can be collated and analysed at the level of individual jobs, so can the cost data, both direct and indirect, incurred within individual work centres. Hence the cost performance of work centres can be analysed and reported by the same module.

There are many forms which these various job costing and work centre performance reports can take. The usual types of reports generated from such a module include the following:

— Job cost sheets and summaries;
— Cost of sales for a nominated period;
— Work in progress valuation at a nominated date;
— Time sheet analysis and history;
— Machine utilisation analysis by work centre;
— Standard cost estimates for jobs (based on standard or synthetic times);
— Historical job cost analysis;
— Parts and material scrap reports;
— Material usage reports;
— Work centre expenses analysis;
— Work centre performance analysis.

The extension of computer aids into the full scope of job costing and work centre performance monitoring can provide timely, and valuable financial and other information. Data on both the work carried out within the company and the work centres within which the work is performed can be accumulated. It can be used to improve future estimating and to identify shortcomings within work centres. However, the quality and value of the information which can be produced from such extensions to the technology is very dependent on the standard of information held and captured by the system and also the speed with which it is acquired.

Conclusion

From the discussion within this chapter it is seen that the full scope of computer aids within the areas of production and materials planning and control is now extremely wide. Whilst each of the specific areas are typically covered by separate application software modules, these may be linked together to avoid duplication of information and to retain cohesive integrity throughout the overall system.

The particular vendor's system determines the basis of mutual dependencies of these application modules. Most systems allow the technology to be extended incrementally, within the set of module dependency rules which apply for the particular system. It is not therefore necessary for a company to commit to full implementation of all the modules at the same time. To do otherwise would be most intimidating for the user company.

Many companies could derive much more benefit from the information which they accumulate within their computer based system in this application area. Often they are not aware of the various reports which are possible from the information held within the system or they fail to impose the disciplines to capture relatively small amounts of additional information required by the system in order to produce further valuable management control information.

So often, senior management tend to stand somewhat aloof from the technology and fail to identify potential benefits of the information which could be provided, particularly on an exception basis. By the same token, the system administrators of such integrated systems may not be fully aware of the potential commercial benefit of the information which the system, for which they are responsible, could produce. The system administrator needs continually to explore the potential for enhancement and tuning of such systems. Both should regularly explore the real commercial needs within the various company functions. The effectiveness of the routine procedures associated with the computer system must be kept under review if the maximum benefits are to be derived from such technology. It is equally important that these often manual procedures operate as efficiently within the company as do the computer modules themselves.

10

Choosing a proprietary turnkey system

The previous chapters have served to identify the plethora of computer based technology aids which are available from a wide range of commercial vendors. A major problem for many small and medium sized companies is the whole business of exploring how this range of technology options relates to their own companies. So often the senior management of such companies are totally pre-occupied with the day to day business of running the company. They are trying either to keep the wolf from the door during the difficult part of the economic cycle or to increase business, and service their markets when times are good.

Often small to medium sized companies do not have staff with specialist skills in the wide range of computer aids from within their own number. Frequently the decision makers within the company are not sufficiently aware of the technology although they have some general level of awareness from trade press and other sources. Such managers hence find difficulty in relating the full potential of the technology to the problems which are either present within their companies or are likely to emerge within the near to medium term future. In truth, most of these companies could derive some benefit from an investment in computer aids for some of the the key functions within the company. The problems are to identify which types of technology package will be of most benefit, from which supplier to purchase, how to structure the company in order to maximise the potential benefits and also preserve the option to extend the technology over time.

As soon as the technology supply community becomes aware that a company is investigating options for upgrading its technology base, its management will be bombarded with propositions to purchase or lease some mix of technology aids. One thing is then certain, as a company manager you will know that this or that technology package will do exactly what is required of it within the company. The reason why this rapidly becomes so clearly known is that the salesman tells you so! All that is required is to bite the bullet, make the decision to invest and then the benefits will start to flow.

Would that such investment decisions were so simple. Unfortunately, there are no easy ways out of the maze of contentious information which is likely to be showered on the company from the technology supply industry. As with any other investment, there is no substitute for a detailed assessment of the needs of the company and how various contending technology packages meet this need. All too often companies which have invested in computer aids have failed to realise the expected benefits. This experience is not only found in small and medium sized companies, who can ill afford to make the wrong investment decisions, but throughout the 1980s and into the 1990s larger companies have needed to re-examine the basis of their earlier technology investments.

So how can a company reduce the risks of either making the wrong investment in computer aids or failing to maximise the potential benefits from the systems they have implemented? Within this chapter the various planned stages of exploring the options for investing in the technology will be discussed. Just as with any other significant investment, a rigorous approach should be taken. Each decision point on the path of introducing and extending the range of computer aids within the company should be part of an overall researched implementation strategy for the company. This strategy should stretch over the medium term future in some detail. Longer term objectives will also need to be identified.

External consultants

Some companies find it advantageous to use the help of an external specialist independent consultancy service. This has the benefit of removing the problem of lack of specialist skills within the company. It is essential to establish that the consultant appointed is truly independent of the vendors of technology packages. It is also preferable if the consultant has a background as a practising engineer rather than one entirely of computer science. The consultant must be able to relate to the company's technical staff and to appreciate the nature of the day to day operation of the key technical functions within the company in a practical manner. It is equally important to identify the areas where computer aids are not as yet appropriate to the operation of the company as it is those areas where benefits could accrue. A good measure of engineering pragmatism should be sought from any consultant who is to be appointed.

Technical review of the company

In any event, the first task in assessing the need for computer aids is to identify the scope within which their potential application should be explored. A series of in-depth technical interviews should be conducted with senior and middle management as well as operational staff within a variety of technical functions within the company. The perceived problem areas within the technical areas of the company should be elicited from these interviews and related to the potential application of computer aids.

It is essential to explore the perceptions of the functions which interface to and from those which, it is thought, would benefit from an upgrade in their technology. For example, if it is perceived by the company management that computer aided draughting would be appropriate to the work of the design drawing office, it is important to explore the perceptions within say design analysis or sales together with those within production planning, test and inspection, and field maintenance. The quality, volume and nature of the information which passes to the function which lie both upstream and downstream of the design drawing office needs to be explored and documented.

So often, if one looks within an operational unit such as the design drawing office, its staff perceive that the information received from upstream functions is invariably less than adequate, subject to frequent changes of variable quality or subject to any number of other deficiencies. The information which such staff pass to the downstream functions such as production, test or maintenance areas, is of good or adequate quality and consistency! By interviewing management and staff from either side of the interfaces with a particular function such aberrations in perception can be in part resolved. The true situation frequently lies between these various perceptions.

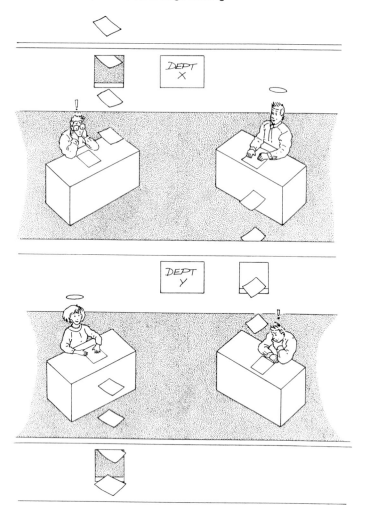

Different perceptions at the interface between technical departments.

During this series of discussions it is important to collect information on the following key aspects of the work which is carried out within the range of operational units:

— The structure of the staff within the unit;
— The main communication channels through which technical and associated data flows within the company;
— The volume of work throughput of the unit as expressed by such parameters as:

- — The volumes and range of sizes of textual documents (e.g. tender documents, specifications, test specifications)
- — The volume and complexity of analysis tasks carried out together with the nature of the analysis
- — The volume, sizes and complexity of design sketches
- — The volume and size range distribution of sub-assembly and assembly drawings
- — The volume, sizes and complexity distribution of non-dimensioned symbolic schematic diagrams (e.g. circuit diagrams, piping and instrumentation diagrams)
- — The volume and size range distribution of dimensioned detail drawings
- — The volume of parts lists
- — The volume and nature of part programs and other process planning documents;
- — The number of revisions which these various forms of documentation go through within their active life time;
- — The structure of work centres, cost centres and profit centres within the company;
- — The volume of factory paperwork (e.g. works orders, job tickets, material requisitions etc) which is typically handled within an average period, together with indications of peak activity;
- — The volume of stock transactions handled within a typical period, together with peak rates;
- — The volume of projects handled by the company during a typical period;
- — The degree to which the work load within the company is effected by seasonal variations;
- — The numbers of stocked and unstocked parts, materials, sub-assemblies and finished goods handled by the company;
- — The breakdown of costs in terms of labour and materials across the various technical cost and profit centres.

It is also vital to identify some form of consensus upon where the bottlenecks lie within the flows of technical information. Similarly, how often does the quality and consistency of technical information generally fall below that considered optimum? Also, the sensitivities to peaks and troughs within the workload within the various operating units need to be identified.

Within a number of the operational units, and particularly within the design functions, it is important to glean information about the patterns of work and the *modus operandi* of the various staff within the unit. For example many senior managers often have at best only an approximate view of how many hours are consumed say within a design or production drawing office. Consequently they can so easily become over optimistic on the potential direct benefits of such computer aids as computer aided design and draughting systems. Some informed estimates of the time spent on the following tasks within, for example, the design or production drawing office should be documented:

— Time spent in discussion with operational units which lie upstream of the functional unit being considered (e.g. for a design drawing office - the time spent in meetings and discussions with say the design engineers, sales and marketing, customers);
— Time spent searching for information (e.g. catalogues, other drawings and specification documents);
— Time spent in design or geometric calculations;
— Time spent in design scheming in the form of design sketches or modelling;
— Time spent in detail draughting (getting lines on to the paper);
— Time spent in general arrangement and sub-assembly draughting;
— Time spent in preparing bills of parts and material, indented parts lists etc;
— Time spent in preparing inspection drawings and templates;
— Time spent discussing with functional units which lie downstream of the unit being considered (e.g. production planning, shop floor, test and inspection, maintenance, on site visits to the customer);
— Time spent in training or administration.

An assessment can be gleaned of where the time goes on tasks not directly concerned with generating technical information. It is also important to estimate what proportion of the time which is actually spent in getting 'lines and text on paper' is spent in modifying information rather than creating new information. It was highlighted earlier that the editing of information which has already been created within a computer system is invariably more cost effective than is the origination of the technical data within the computer system in many cases.

In the case of the engineering drawings of the various types which have been listed above, it is important to make some form of assessment of frequencies which the characteristic features favouring the use of computer aids are to be found. These features were discussed earlier within Chapter 7. It can be of some help to collate a set of sketches, arrangement and detail drawings which have previously been generated for one or more typical products or projects within the company. A formal score can be recorded for each of the drawings for its content of each of the main characteristics which favour the use of computer aided techniques.

From analysing the information collected and documented (as a result of discussions identified above) a view can be formed as to which main function within the full scope of computer aided technology is likely to be of benefit to the company. This assessment should be based on the short, medium and longer term development of the company. Priorities can be established and the basis of a strategy for the introduction, and further extension, of computer aided techniques can be developed. This strategy should be agreed within the senior management of the company. A plan for evolving the technology within the company over time can thus be postulated and the target benefits, to be sought, documented against this plan.

Functional specification

The next task is to identify the technology packages and modules to be explored further, from the company's strategic plan. These should cover the short and medium term needs. Then preparation of a document specifying the functional scope and required capabilities of each of the technology packages, should be undertaken. This document is known as the functional specification for the technology package(s). At the top level it should identify, in engineering and operational terms, the scope of the modules planned to be implemented. It should relate to the application background of the company, against which the computer aided system(s) will need to operate successfully.

Much will depend on the skills available within the company as to whether this task can be properly entrusted to a member of the in-house staff or to an outside, fully independent consultancy organisation. It is important that the functional specification document identifies the key functions, command structures and preferred *modus operandi* which is sought by the company. It must also place the functional operation of various computer aid modules in the context of the operational loads which, it is envisaged, will be placed on the system. Some provision should be made within the functional specification for projected growth, both in the breadth of operation of the technology and volumes and nature of data over later years of operation.

Future links and interfaces with other technology modules should be indicated within this document. Also, a statement of requirements should be included with regard to such operational considerations as worst case system response times, data field sizes, flexibility of report generation, system housekeeping needs etc. It is important that these target performance characteristics are maintained into the future, when the volume of information may well have grown as the system absorbs operational company data over time. All too often, a supplier will quote for an initial set of hardware which, whilst supporting initial data demands with adequate response times, may not be adequate for future company loads.

The compilation of the functional specification should be the responsibility of a nominated member of an assessment team. He or she should ensure that key members of the company are aware of its contents and their views are adequately represented. A consensus should be established in preparing the document. It may be that in preparing the functional specification, sufficient shortcomings are identified within existing company methods and procedures, that a computer based approach would stand little chance of being used successfully. Computer aids will only benefit the company if required information can be identified and communicated efficiently within the company. Whilst computer aids assist in imposing good working disciplines, they do not offer a substitute for basic organisation and operational integrity. In such cases, the attempt to specify how computer aids would help the company is not entirely wasted. If the pre-cursors to their successful deployment are identified as missing this in itself is valuable information. These organisational problems can then be solved with a plan to review the potential for wider use of computers again at some later time in the medium term future.

Assuming that the use of computer aids still looks an attractive proposition, the following steps need to be taken. The final functional specification document should be written in such a manner that it can provide a good basis against which candidate vendors can configure and bid their proprietary offerings. It should also form an important document within the set of contract documents which covers the eventual supply of the technology package(s). There may well need to be agreed schedules to the supply contract which identify exceptions and concessions relating to the original specification document. There is no substitute for a formal statement of requirement document and the functional specification should serve to meet this need.

Hopefully, any problems which occur in the future use of computer based systems, will be resolved through a good working relationship between the supplier and the company. In some sense, if the company has to fall back on the formal contractual relationship which exists with its supplier, this would indicate something has gone wrong. However the company must protect its contractual position with a comprehensive functional specification of its needs.

Seeking vendor responses

Having identified the focus of company's needs and documented them in a functional specification, a list of candidate vendors should be identified. A search of specialist trade press, exhibition catalogues and conference proceedings will serve to identify vendors of technology packages focused on the market sectors within which the company's needs fall. Alternatively, if an external consultant is helping the company, he or she should have a good working knowledge of the most appropriate vendors. Vendors within any particular application area can be considered as belonging to one of the following three general types:

— Software vendors and developers;
— Predominantly hardware vendors who also license software packages;
— Turnkey system vendors.

Software vendors

The software vendor community develops application software packages or market packages under license from the owners. They may also be in the business of developing software solutions on a custom-made basis for their client companies. Generally, they license software into a market which has already installed computer hardware and associated equipment. In doing so, they identify the minimum recommended hardware configuration needed to allow their software to run in a satisfactory manner. Such vendors will sometimes offer a service to customise either standard software packages or custom-made software systems to meet the particular needs of the company. For the more advanced software packages, they will normally offer training on the application software and a full software support and upgrade service against an annual fee payable to vendor. This fee is typically based on a percentage (typically 10 to 12%) of the list price value of the single payment licence fee for the software which is installed. At the lower cost end of the industry some software vendors operate through mail order, offering standard application software packages. Training in the use of such lower cost software systems tends either to be left to the user company, working through the training or operations manual provided with the package, or through third party training organisations.

Hardware vendors

Most hardware vendors see their main business as centred around selling and supporting hardware installations. They are sometimes known within the industry by the term 'box shifters'. Their main motivation is to sell boxes of hardware, but in doing so recognise that software products need to be offered to their client companies. They will sometimes work in a loose collaboration with a software vending company and will jointly bid for business. Again, at the lower cost end of the hardware market, competition is very fierce and much equipment is sold via mail order or through high street outlets.

Turnkey system vendors

The so-called turnkey system vendors offer a more comprehensive supply, training and support service, from a single source as far as the client company is concerned. They are called turnkey suppliers because, in theory at least (and frequently in practice), the full hardware and software system is supplied from a common source. The 'key' is turned on and the system operates as envisaged by the supplier straight away.

Turnkey vendors will typically offer application systems on a somewhat restricted number of hardware manufacturer's equipment, in order to limit their future support obligations. From within this small range of hardware manufacturers, they will select the most appropriate equipment to support the application software. They make their profits from discounts received on the hardware they sell, the software licences sold and training and annual service fees. Such vendors tend to charge a higher price for the hardware content of the system than could be achieved from a volume supplier of hardware. They tend to work to list prices for both hardware and software licence fees, or a relatively small discount on list price, depending on the prevailing market conditions. However, they tend to offer a more comprehensive service to clients, in terms of training in the use and administration of the combined hardware and software system.

For many user companies there is an advantage in having a single point of contact and responsibility for both hardware and software support. This avoids any contention in agreement of the source of any ensuing problem. Such difficulties may arise if the system's hardware and software components have been supplied by different vendors. It may well be that the turnkey system vendor sub-contracts some aspects of support or training to other parties. It does however still act as a common point of contact for the user company. This can offer advantages, particularly when the system has been fully integrated into the operation of the company. Any problems which do arise need to be remedied quickly if effective and efficient operation of the company is not to be jeopardised. The possible extra capital cost often implied in buying from a turnkey vendor may be recovered over time from having this clear unequivocal position with regard to system support.

It is important to choose the list of potential system suppliers with these wider objectives in mind. Initial enquiries with candidate vendor companies should serve to clarify what the nature of the supply companies are. It should also confirm that they are appropriate to be included within the initial survey of suppliers. Also, from these initial contacts with potential candidate vendors, some budgetary cost information can be gleaned which can help to refine the company's initial view on necessary capital expenditure.

Having chosen the so-called 'long list' of candidate vendors, each vendor should be invited to respond to the agreed functional specification for the computer aided system(s). A copy of the functional specification, together with a covering letter should be sent simultaneously to each vendor on the 'long list'. Within the covering letter the following information which is sought from the vendor should be identified:

— A non-compliance document identifying which of the functions within the functional specification is not supported by the vendor's system or which is supported in another way within their system. Functions provided in a different manner should be discussed within the document;

— A list of functions which will only be supported within a planned and committed future release of the vendor's application software;

— An itemised budgetary quotation for the proposed recommended hardware content of the system. The hardware should be specified by the candidate vendor, having regard to the projected volumes of information to be processed within the system and the required target system response times - both of which should be identified within the functional specification. This should identify budgetary costs for the following classes of hardware equipment:

   — Computer processors and their types
   — Hard disks (with speed and capacities)
   — Access terminals, work stations etc
   — Printers
   — Plotters
   — Other computer peripheral devices
   — Communication equipment and any bearer circuit line connection charges;

- An itemised budgetary quotation for the software modules to be supplied and any customised software required to meet demands of the functional specification. These should itemise the following:
  - Operating system license fees
  - Communication software license fees (e.g. networks)
  - Application software modules licence fees
  - Software customisation fees;
- Budgetary costs for the following pre-acceptance services:
  - All delivery charges
  - Any required site survey fees
  - All installation wiring, cabling and other site works
  - All commissioning and acceptance testing work
  - Any pre-delivery project, implementation or account management fees
  - Starter pack of consumables (e.g. special stationery, removable disks, tapes);
- Recommended training course fees;
- Recommended spares to be held by the company and their budgetary costs;
- Budgetary costs of the ongoing annual hardware and software support contract fees for the hardware and software which they are proposing;
- Budgetary estimate for the annual costs of consumables to meet projected usage of system as indicated within the functional specification;
- A set of their latest published accounts;
- A narrative identifying experience of the vendor in supplying similar systems within the market sector within which the company most naturally can be classified. This should include a list of appropriate firms to which the vendor has supplied similar systems together with the years in which they were supplied;
- Information on any user groups which have been set up and are active with reference to the application software which is being proposed.

The advantage of requesting a non-compliance document from the vendor is it serves to concentrate on the area where the vendor either does not meet the functionality requested within the functional specification or claims to meet the objective using a somewhat different approach. If several vendors have been approached, as is most prudent at these early stages of enquiry, and each vendor responds with great volumes of information which only serve to confirm that they are able to meet each of the points covered in the functional specification, this can be extremely tedious for the assessment team. It is far better to attempt to identify where there are differences from the functionality set out in the functional specification. Taking this approach can minimise the volume of information to be assessed whilst not prejudicing the important aspects of what is being proposed. It does however emphasise the importance of preparing a well considered functional specification document.

Within the covering letter to candidate vendors, it should be made clear that the functional specification, with agreed exceptions and concessions, will eventually form a part of the contract of supply for the proposed system. All of the candidate vendors should be requested to submit their responses by a given date, typically three to four weeks from the date of sending the invitation to them.

On receipt of their responses, the person responsible for co-ordinating the preparation of the functional specification should analyse the responses. He or she should prepare an internal report which identifies the main strengths, weaknesses and budgetary capital and revenue costs for each of the 'long list' of candidate system vendors. This report should be prepared objectively and carefully in order to distil the volume of paperwork sent in by the candidate vendors. The report should be circulated to each of the key interested people within the company prior to holding a formal selection meeting to identify a final short list of vendors.

At this crucial meeting some of the vendors will be eliminated, either on the basis of budgetary costs or on more than acceptable non-compliance with the functional specification. The criteria for assessment of the vendors should of course be agreed amongst the assessment team and documented, preferably prior to the key assessment meeting. Often the criteria have not been agreed. In such cases they must be the number one item on the agenda. The weighting of the factors to be used within these criteria should be explored and a consensus sought amongst the assessment team as to what the priorities are to be.

Following what will often be a somewhat brainstorming assessment meeting, it is likely that more than one of the vendors will appear to be worthy of further detailed examination. Typically, a final short list of about three vendors is likely. The next stage in the selection process should be to conduct so-called 'scripted benchmark tests' at each of the vendor's premises.

Benchmark testing

The purpose of the benchmark tests is to evaluate the vendor's system suitability for the mix of work carried out by the company. Each shortlisted candidate should be invited to undertake the same benchmark tests, against a prepared script. This arrangement provides an objective basis of comparison of the candidate systems. The series of tests can eventually lead to final system selection. Such tests may involve identifying limits of the system's capabilities but the primary objective is not to cause the system to fail but to test its ability to meet the company's needs.

Scripts

The reason for preparing a script is to ensure the company retains a good measure of control over the conduct of presentation of the vendor's system. In fact, two scripts should be prepared. The first for guidance of the vendor, and the second for use of the company's assessment team during the visit. The two will be similar in many ways, but the version to be used by the assessment team should be considered more as a structured check list. It should make provision for each team member to record their objective and subjective views and measurements, resulting from their observation of each task and test requested of the vendor.

A fresh copy of the checklist should be used for each vendor visit. The assessment team should record their name and that of the vendor on their copy of the script for future reference. After only a small number of visits to different vendors, it is extremely difficult to recall from memory details of the results of any one test. Hence, the importance of each assessment team member recording their views and measurements during the visit in a structured manner.

The benchmark test scripts should set out the sequence of inspections and tests which need to be carried out during the visit to the vendor. They can be written to generally follow the sequence of topics covered within the functional specification. They will normally need to be read in conjunction with the functional specification, in order to place the various tests in the overall context of the company's needs.

A convenient starting point for the visit is to inspect samples of the user and system documentation for the proposed application software modules. The quality of the system and user documentation supplied with the system is important to the company over the system's lifetime. A user of computer aided technology must be able to understand and use the documentation as a first level of system support, subsequent to appropriate training. Confusing, inadequate, inaccurate or out of date documentation would represent a serious weakness of the system.

It may be appropriate to assign a subset of the assessment team to inspect the documentation in some detail. This can be done to advantage, typically during the time that the vendor's sales staff are making their initial sales pitch and telling the rest of the team how their company is able to supply all the solutions to the company's needs, at a very competitive cost, and at very low technical risk! The documentation should be assessed against criteria important to its ease of use within the company. The points to look out for and questions to ask of the documentation include:

— Is the documentation clearly presented and well structured?
— Is the documentation up to date?
— Is there a clear index to key operational topics?
— Is the structure highly cross referenced, such that the user has to swap about a lot during researching a functional operational sequence of system commands?
— Are clear and relevant worked examples included?
— Are sample screen layouts shown together with the examples?
— Are short form summary cards or documents provided for quick reference for the experienced user?
— Is the documentation robust enough for day to day operation, or will it fall apart after two months?

Having made an objective assessment of the documentation, it is vital to enquire into the nature of hardware upon which the system is to be demonstrated. It is important that wherever possible, the vendor should present the benchmark tests on the hardware which they propose to offer in their tender for supply. This may prove to be difficult in some cases, but it is important to record the basic hardware configuration which is to be used for the benchmark tests. It may be necessary to repeat some of the more response time sensitive tests on other hardware, closer to that being proposed - possibly at the site of an existing operational user of the system. Options on terminal and workstation types, printers and plotters and user interface options should also be summarised and explored. Observations and perceptions with regard to these options should be recorded during some part of the meeting.

The benchmark script should then set out the various tasks which the vendor's demonstrator staff will be asked to perform. The nature of these tasks will depend on the type of computer aided application which is under consideration within the tests. By way of an example, if computer aided draughting is to be assessed, a series of tasks should be scripted which relate to the typical design and draughting operations which are frequently encountered within the company. Clearly, the sequence of draughting tasks needs to be constrained to what it is reasonable to be undertaken within the time set aside for the benchmark tests. They should however cover the sequences of design and draughting operations which are met in the day to day work of the company, for example, the functionality of the system for design sketching, geometry construction, geometry manipulation, detail and assembly drawing draughting, hatching and dimensioning, annotation and parts listing.

Similarly, in the case of computer aids for draughting, other tasks could also cover such aspects as deriving parametric components, handling component libraries, construction and manipulation of various forms of three dimensional models and interfacing to other modules such as numerical control part programming. The criteria for good functional practice within computer aids have been discussed earlier within this book. These and other criteria, determined by the company, should be assessed and the results recorded by each member of the assessment team, for example, the desirability of associative dimensioning and hatching within computer aided draughting systems.

Demonstration

At the end of each section of benchmark testing, the vendors should be offered the opportunity to highlight any particular features of their system which they feel is not adequately identified within the functional specification. Care should be taken to ensure that the vendor is not allowed to steal the initiative from the company and go into a long series of well rehearsed standard demonstrations. These can be selected to look impressive, rather than to address the identified needs of the company. Many experienced demonstrators of computer aided systems are well versed in presenting their system in the best light and successfully covering up its potential or actual deficiencies.

One problem with any demonstration of a computer system, is the observer can become just as impressed with the technical performance of the well rehearsed person demonstrating the system, as they are with the system itself. This may subjectively colour the judgement of the less mature members of the assessment team. Hence it is important that the assessment team retain the initiative during tests. The demonstrator should be largely constrained to cover the key areas of functionality of the system which the assessment team (sometimes with external help) has identified as being of greatest importance to its use within the company. This has to be balanced with the need to allow the vendor to show off facilities of the system, which in their view, are not adequately covered within the tasks set out in the benchmark tests. These additional areas can be covered as a separate activity after the tasks set down within the script have been completed.

It may be, that in order to save time during the visit, the vendor suggests that some of the work of creating information covered within the tasks, is undertaken prior to the visit. Whilst this may be appropriate in some cases, it is important for the assessment team to satisfy themselves not only that the task can be successfully completed using the system, but that methods used and completion time are also acceptable. In agreeing to work being undertaken by the vendor prior to the benchmark test visit, it should be stressed to the vendor that methodology used must be shown during the visit. This can be particularly important for topics like handling parameterised geometry, within a draughting or modelling system, or automatic printed circuit board routing where the method is at least just as important as the final result.

At the conclusion of the formal benchmark tests, a number of other topics should be discussed with the vendor, in order to form a basis of comparison with the other shortlisted vendors. The following topics should be included:

— Support arrangement, covering such topics as:
  — The location from which both hardware and software support will be provided
  — The number of hardware and software support staff available within the vendor organisation
  — The availability of direct on-line support from the vendor's site, using modem facilities - in which the vendor can dial into the company's system to explore reported problems
  — Guaranteed response time for support
  — Fault reporting procedures
  — The frequency of software updates and upgrade frequencies. If the software is currently being updated frequently (say more frequently than every six months or so) this may be a sign that the software is rather new or lacks maturity or stability;
— Training arrangements, covering the following aspects:
  — The recommended training periods for different modules within the system
  — Is training to be carried out on site at the company or vendor's site?
  — The availability of free training credits with the initial supply of the system
  — The recommended sequence of training of company staff;
— Transfer arrangements for data which may exist within any existing computer based system within the company;
— Potential upgrade paths from the system which is being proposed by the vendor;

— Proposed customisation of the vendor's system;
— Compatibility with data transfer standards;
— Proposed arrangements and timescales for delivery, installation and commissioning;
— User group activities;
— User site reference visits.

The benchmark test visit to a vendor will be a very busy period. The same assessment team should make every effort to attend each of the visits to the shortlisted vendors, in order that they can collectively compare views on strengths and weaknesses of the candidates. It is often wise to include one member of the company staff who is likely to act as a devil's advocate. Such a team member will help to balance the views of enthusiasts for the technology who tend to self select themselves into the assessment team.

Following visits to the vendor, further visits to users of the shortlisted systems should be strongly considered. There is no substitute for discussing the use of the contending systems with practical users of the systems. Exploring their experience of both the operation of the system and the relationship which they have had with the vendor can be of benefit. The vendor should be asked to arrange a visit to one of their existing users whose business and use of system most closely matches that proposed for the company. Ideally, the proposed user company should be using the application software on the same hardware configuration which is being proposed by the vendor.

During the visit to the user site, any technical areas which have been identified as worthy of further study should be explored in more detail. The various topics which were identified above concerning the relationship with the vendor should also be explored with the user site staff and their observations recorded.

The team will be faced with a great deal of new information during the various visits. If they fail to record their observations and perceptions on the benchmark scripts provided, and also during visits to the vendor sites, they will become confused. They need to be able to assess which system did what well, and which poorly at some time after the visits. During the visits, one person from within the assessment team should be appointed to ensure that tasks are completed and all topics of interest covered. It is essential that the meetings do not get bogged down in detailed discussion within any one narrow topic to the exclusion of others.

When all the benchmark tests and user site visits have been completed, each of the shortlisted vendors should be invited to prepare their final quotation. It should be based on their best prices and each candidate should quote against the same refined specification. The quotation should be itemised and cover all aspects of capital and revenue costs associated with implementation of the system within the company. The vendors should be asked to respond by the same date.

Final vendor selection and contract considerations

Upon receipt of the final quotations, the assessment team will be equipped with all the information needed to complete their final selection. The criteria for final selection should be reviewed and agreed amongst the team. If the services of an external independent consultant are used, his or her role will be to provide technical guidance and interpretation of the information gathered during the selection process. The consultant will have attended the benchmark test visits as a part of the assessment team. He or she will have recorded observations based on his/her specialist experience. However the final decision can only be vested in the assessment team members who are employed by the company. The company must make a success of the use of the system within its operations and its members alone must make the final decision on which system to acquire.

Sometimes, it will be seen that more than one vendor, on balance, can provide a technical solution which would be satisfactory to the company. There are likely to be some variations in the perceived technical performance between the different systems but it is often the case that these may balance out across the whole scope of the system. In such circumstances the sensitivity of the final decision to the technical issues is reduced. The final selection will depend on normal commercial considerations of price, stability of the supplier, support arrangements, perception of the quality of the vendor's staff and even a basic gut feel about the vendor. This situation can be encouraging to the company embarking on a substantial investment in upgrading its technology base. It is, however, difficult to be sure that this is the case without undertaking the extensive technical assessment which has been outlined within this chapter.

Contract of supply

Having made the provisional final selection, it remains to negotiate the contract of supply with the chosen vendor. Most vendors will attempt to supply the system under their standard terms and conditions. These cover supply, installation and commissioning, training, and support and maintenance. They will typically cover both hardware purchase or leasing and software licensing arrangements. It is preferable, for a company which is making a substantial capital investment, to ensure that their functional specification document, in which are contained performance parameters such as operational response times and data volumes is included within the supply contract documentation. Agreed exceptions and concessions to the functional specification can be included in a separate schedule within the contract documentation. Other schedules, covering promised extensions and customisation of the system to meet the company's needs, should be agreed with the vendor and also included.

Customisation and phased implementation

Where the system is to be supplied with extensions and customisation to the standard software, a part of the fee for software should preferably be held in escrow until extension and customisation work has been fully tested and accepted. An appropriate series of acceptance tests for this work should be agreed with the vendor. This is particularly important where it is necessary for software extensions or customised routines to work in an integrated manner with the rest of the standard system. This aspect must be fully covered within the acceptance tests.

In some cases, particularly in the application areas of material and production planning and control, a phased implementation of the various modules will be needed over a period of time. For these application areas, it is unlikely that all of the modules could reasonably be introduced into the working practice of the company at the same time. A planned phased introduction of the technology should be agreed with the vendor. They will give guidance on the mutual dependencies of the various modules and sequence of introduction of the software packages. It should be possible to agree phased licensing terms for the total set of modules appropriate to operation of the system within the company. There may be time limits imposed on the period of the phased implementation by the vendor which relate to any special price terms which apply.

For some company applications, substantial software customisation may be needed, in the form of entirely new software modules. For example, a new module may be needed for a specific design analysis or simulation routine. It is important to agree with the vendor the status of the new software module. The ownership of the software module needs to be agreed, together with any arrangements which will be made for any subsequent commercial exploitation of it. In such cases, if the full commercial cost of the development of the new software module is to be paid for by the company, it is reasonable for the company to have the source language version of the software delivered as part of the contract of supply. There may be problems which need resolving, where the customised module makes use of other software modules which are the proprietary property of the vendor or a third party. These will need to be considered and agreed. Such agreements and arrangements are far better addressed and resolved prior to signing the contract of supply.

The computer system supply industry has been subject to cyclical periods of growth, success and expansion countered by downturn, contraction and rationalisation. As with any new and rapidly developing industry, there has been a series of new companies, mergers, company failures and take-overs. The users of the technology need to consider their own protection in the event that the vendor, at some time in the future, can no longer provide the appropriate level of support or they may disappear from the market altogether. There is no fully satisfactory solution to this potential problem. Successful support, particularly of advanced and complex engineering related software, is greatly governed by the continuity of the services of highly skilled and specialist software and engineering application support staff.

The software which is delivered by a vendor under a software licensing contract, is often in the form of executable code. That is to say, the version which the company has on its premises cannot form the basis of independent software support if the supply company goes out of business or is no longer able to offer software support services. In order to support software, to resolve errors within it and to develop it further, it is necessary to have access to the source code version of the software. This is the version of the software in which it was written in some high level programming language.

As a safeguard against any future failure of a single source vendor to be able to support its software, the vendor can be asked to place a copy of the source code in a bank from which, in that event, the client company can gain access to it. This does not provide a total protection against future failure of the vendor to meet its contractual obligations with regard to the support of the software. The client company is unlikely to have the skills to use the source code of the software in order to resolve operational problems within it. However if the company preserves its longer term access to the source code it could seek software skills elsewhere.

This procedure only has value where the capital value of the software is substantial, or where failure by the vendor to support or to further develop the software could be prejudicial to the client company. It does however afford some protection against the commercial failure of the vendor or the vendor's decision to withdraw from the market.

Planning the implementation

Having completed commercial arrangements for the supply of the computer based system, it remains to plan for the implementation. The internal planning for the introduction of a substantial computer aided system needs to be thorough. Assistance should be sought from the chosen vendor of the system. They will recommend certain tasks, which the company needs to complete prior to the delivery, installation and commissioning of the system.

Equipment siting

Consideration needs to be given to the siting of the various elements of the hardware of the system. Any modifications to the building structure of the area where the equipment is to be housed should be planned well ahead and implemented during the lead time for the equipment. Fortunately, the physical size of modern computer hardware is relatively small and modern computers are tolerant of normal office environmental conditions. However, it is wise to ensure that the location for the equipment is generally clean, relatively dust free and designated as a non-smoking area.

Whilst full air-conditioning equipment is not normally required consideration should be given to avoiding extremes of temperature and humidity in areas to be used for housing equipment. Workstations can emit significant amounts of heat during their day-long operation. This, together with excessive solar input through large expanses of windows, can result in uncomfortable working conditions during the summer months unless planned for.

The siting of computer terminals, workstations, and the other peripheral equipment such as printers and plotters should be planned prior to the delivery of the system. If peripheral devices are to be connected directly to the computer processor (rather than via a local area network) there are limitations on the maximum cable distance between the computer processor and each of the devices. If the installation is to include local area networks, consideration must be given to the routing of the network cabling. It is wise to provide for additional access points on the network to cover for possible future additional network nodes or the re-siting of terminal nodes. Some types of printers and plotters can be somewhat noisy in operation, and care should be taken as to where to site them. A compromise often has to be struck, in that they need to be conveniently placed for frequent access, but not intrusive on the office working environment because of their noise level.

A storage area needs to be planned nearby for holding computer consumable materials. These include stationery, off-line magnetic storage media such as magnetic tape cartridges and floppy disks and printer ribbons. As will be discussed later within this chapter, there is a need to hold security back-up copies of both software and operational data. Similarly, archived data, removed from the on-line data held on the disk units will need to be stored on off-line magnetic media. These off-line program and data files are essential to the company in the event of any future damage or serious equipment failure. It is therefore important that they are kept in a protected area. It is recommended that these copies are held in an area away from the computer system, preferably in a separate building and under fireproof conditions.

Electrical considerations

Within many industrial premises, and notably within industrial estate locations, stability of the electrical mains supply cannot be fully guaranteed. Transient spikes and surges in the mains supply, due to operation of industrial equipment (e.g. electrical welding equipment), can cause problems to high speed electronics equipment such as computer and associated communications equipment. Invariably, the system vendor will strongly recommend that a mains protection unit is included in the mains power supplies to the system's electronic equipment. The effect of such a unit is to stabilise the mains supply to the equipment, filter out mains spikes and ensure that, in the event of the loss of mains supply, the computer equipment shuts down in an orderly manner. The various electronic and electromechanical subsystems can thus be protected and loss of information minimised.

Power supply cables and data cable runs need to be planned and agreed in liaison with the chosen system vendor. Normal segregation rules for such cables need to be followed. Particular care needs to be taken in the siting of both cable runs and equipment within the shop floor area, particularly where the area is subject to frequent traffic of heavy equipment. Reducing the risk of accidental damage to both cabling and equipment in workshop environments is a self evident requirement.

Where the system is to operate between more than one physical location and data communication circuits are needed, these will be specified by the system vendor. The vendor will normally include supply of appropriate communications hardware required at either end of the communication links. The bearer circuits between sites will normally be supplied by one of the national telecommunication companies. The type of bearer circuit will be recommended by the system vendor. A proper choice of bearer circuit is essential in order to service the data transmission rates which must be supported, if satisfactory system performance is to be assured. It is therefore important to identify the needs for the communication links at an early time and seek their installation from the national supplier, prior to the delivery of the system equipment.

Planning early system use

Having made plans for the physical aspects of the new computer based system, attention should turn to the strategic planning of the use of the computer aided system. It is important that some beneficial results are derived from the investment as soon as practical if confidence in the technology is not to suffer within the company. The system will often impact on more than one function within the company. It is necessary to involve these functions at an early stage in realising the potential benefits. They need to work together in agreeing the initial objectives for the new system and to plan the staff and other resources to meet these objectives.

Staff considerations and training

The first essential is to identify the staff which will be initially trained in the use of the system. Generally, younger staff tend to be more responsive to new methods but this is by no means exclusive. Certainly the younger staff are likely to be more familiar with basic ideas associated with computer based technology. Most of them will have had some exposure to computers within their professional and technical training. Virtually all engineering related educational establishments now include some training and exposure to the use of computers, although the depth of exposure is still patchy. If it is possible, selecting a mixture of younger technology enthusiasts and older, more experienced staff, who have a good understanding of the work of the company would be the most satisfactory approach.

If a company has a formal relationship with a trade union organisation, it is prudent planning for appropriate union members to be involved within the planning process. They should be included, where indicated, within the initial batch of staff to be formally trained. The attitudes of the trade union movement to the introduction of new technology aids are generally constructive. It has been widely accepted by labour organisations that the acquisition of new skills is of benefit to members and to the overall competitiveness of the company. Both of these benefits are in the interest of the union members and to the union itself. Occasional scare stories of a Luddite attitude from the representatives of organised labour tend to be exaggerated, particularly in the small to medium sized companies within manufacturing industry. That is not to say that there may not be pressure for the recognition of the new skills acquired by staff in their remuneration.

Formal training with the vendor or his or her nominated training agency needs to be planned and scheduled. The initial training sessions should be completed shortly before the planned installation and commissioning of the system. This approach ensures that trained staff can reinforce their knowledge of the system, derived from their formal training, by continued early use of the system after installation. This initial core of trained staff will need to serve as the spearhead for the technology. They will need to show the benefits of the investment and to act as the internal selling agents for the new methods, particularly to some of the more sceptical staff. Future training of more staff can be undertaken by a mixture of more formal training from the system vendor, and internal training by the initial batch of trained staff.

Developing the use of the system

In planning the application of the new system, once initial staff have been trained, emphasis should be on performing a series of tasks which are at the simple end of the spectrum of operations of the company. Much will depend on the scope of the system which is being installed. For example, if the system is a computer aided draughting system, destined for the drawing office, consideration must be given to identifying the straight forward tasks for which the system will be first targeted. It would be foolish to face the recently trained staff with the most difficult and complex tasks at the outset. In this case early emphasis should be on acquisition or creation of libraries of drawings of components and relevant assemblies. If parametric libraries are appropriate, a subset of the trained staff will need to be trained to this more advanced level of competence.

In the case of design analysis or computer aided draughting applications, it is often helpful to spend some period attempting to perform retrospective work on the system. This would serve to familiarise the users with the operation of the system. It would also eliminate risk of interrupting the progress of a live project. Such retrospective tasks can help in assessing the areas of operational work on which to deploy the system in the short term - in order to demonstrate benefits at an early date.

For computer applications covering various aspects of materials and production planning and control, the sequence of implementation and operational use of the software modules needs detailed planning, with the help of the system vendor. Within such highly modular systems there are dependencies between, not only the modules which must be implemented, but also the minimum set of associated data fields which must be used within these modules. It is important to think through the phased sequence of objectives being sought. From training sessions and detailed discussions with the vendor, a list of tasks which need to be completed outside the computer system can be identified. These need to be tackled prior to full implementation of individual or sets of application software modules.

Some of the tasks implicit in moving to the more disciplined world of computer systems in areas of planning, scheduling and controlling of parts and materials and production work centres are listed below. These must be accomplished before effective benefits can be gained from the technology.

— Design and implementation of a part numbering system for the company;
— Collation of basic information relating to component and assembly part numbers:
    — Specification information
    — Cost information
    — Supplier information for bought out items;

— A definitive understanding of the structured (indented) bills of parts and materials for products within the company;
— Collation of stock control information relating to a part number:
    — Stock locations for part numbers
    — Stock control parameters for the part number
    — Alternative part numbers;
— Identification of work centres with:
    — Plant and equipment resources (including operating parameters)
    — Schedules of labour resources
    — Labour rates and overhead recovery rates
    — Rates for capital plant;
— Collation of manufacturing synthetics for standard parts and sub-assemblies;
— Collation of manufacturing routing information;
— Collation of master file information for sales customers;
— Collation of master file information for suppliers.

Clearly, the above list of tasks represents a substantial workload which cannot all be achieved at the outset. The sequences of these tasks need to be related to the proposed implementation of the various application modules. For many small and medium sized companies, some form of basic manual systems will be in place prior to the introduction of computer aided techniques. These manual systems need to be examined in detail, to identify accurate sources of information. This information will need to be input into the master files and databases of the new computer system. Missing information can then be identified and researched in good time to provide for the planned implementation schedule.

In some cases, there may already exist some basic subset of a computer system previously provided from a different supplier. It may be possible to transfer some of the master file information from the old system to the new one. The new system vendor should be asked to explore the extraction of data from the old system master files, and its translation into an appropriate format for the new system. This is only worthwhile if the volumes of data are substantial and the data is known to be of good accuracy and quality. Much will depend on the precise situation at the time when the new system is installed. Often it is found, on balance, preferable to load the new master information from scratch.

In many smaller companies, particularly those growing quickly from the stage where information has been conveyed by word of mouth, or by notes between key individuals, basic information is missing. This should have become apparent in the initial study. It would be folly to have pressed ahead with plans for extensive computerisation of planning and scheduling of manufacturing information for example, if so much of the basic control information and structures have not been previously identified and documented. In such circumstances, a period of operating is needed with a more formal manual system, within which basic company master information can be built up. During this interim period new operational procedures and disciplines, can be introduced. Such an intermediary system can be designed to mimic, to some degree, the operational control procedures of a computer aided system - albeit probably at a lower level of detail and efficiency. It is also possible to include such simple computer aids as standard computer based spread sheets to help in such activities as time scheduling, monitoring and cost analysis. Only after such companies have adjusted to more formal methods of handling the technical information which is flowing within them, should the move be made to using integrated computer aided systems within these application areas.

System administration

In introducing a new computer system to a company, consideration has to be given, not only to its operational use within the day to day functions of the company, but to the administration of the system itself. The storage of associated consumable materials was mentioned earlier in this chapter. These materials need to be monitored and controlled just as any other materials used within the company. There is also a need to identify a person within the company to attend to the management and administration of the computer system itself. This person will need to be trained in the use of the administrative routines of the system and will have routine tasks to perform at various regular intervals.

The system administrator will need to be trained to some minimum level in the use of the computer's operating system. He or she will need to control and monitor the housekeeping routines needed to keep the system in a secure and tidy condition. The administrator will normally provide the point of contact within the company with which the vendor will interface. This will be the link for future support and updating of the system, both in terms of hardware enhancements and software upgrades.

Access to both the functions of the system's application software and to company data evolving within the computer system needs to be controlled. Different operational staff will need different levels of access to the various types of data. They will need to be constrained as to the functions to which they will have access. For example, modifications to an authorised and released company drawing should be constrained to appropriate staff, although a wider group of staff should be capable of calling up the drawing on their screen to view its contents. Similarly, the creation of a new supplier master record on a purchase order processing system will often need to be controlled, or the display of sensitive cost information restricted to selected staff who need to know.

Within the system administrative routines there are often facilities for setting up access rights for nominated users. These rights include access to certain system menu options and the right to view or selectively modify data files. These facilities need to be understood by the system administrator. It is also necessary to identify the needs of each of the various users within the company and to assign the various access rights needed to perform their operational tasks.

The system administrator needs to consider the file organisation facilities supported by the system's operating system. Both software and data files are typically held within a hierarchy of so-called 'directories' within the on-line disk units of the computer. A logical structure of file directories associated with the application software modules, and the associated user data files, is essential if information is not to be accidentally lost, overwritten or mislaid within the computer's disk memory.

It should be the responsibility of the system administrator to set up, monitor and control the main structure of file directories within the system. He or she must set up the appropriate controls to allow the required access routes to this information for the various system users. The system administrator needs to identify normal physical storage devices within the hardware system upon which these file structures are to be held in their master form. If the system contains a local area network, the master files will normally be resident on the file server node disk unit for example.

Parallel running

Where any computer system is to replace a pre-existing method of controlling and monitoring technical information within the company, it is important that the new system is seen to be producing information compatible with the old system. (It may of course be producing rather more information than the earlier system. It is also likely that the information will be produced more rapidly.) In order to feel confident that the new system is performing in a satisfactory manner, a period (approximately two–three months) of parallel running with the old and new systems is mandatory. Only when the new system is running consistently with the old one should the old system be discontinued.

Back-up and error recovery

Within a relatively short period of time, following the introduction of the new computer based system, the value of the company information held within the computer will become significant. Accidental loss or corruption of this data would have a highly adverse effect on the company's operation and could lead to major instability in the company's ability to service its market. It is therefore necessary to implement a rigorous data protection policy from the outset. This consists of regular copying of live company data to back-up copies held off-line in a secure, preferably fire protected, storage location.

The equipment used to store on-line data normally consists of fixed magnetic disk units. These are essentially electromechanical devices, albeit with a substantial electronics content. They are required to work over long periods within tight tolerances. These subsystems can fail and the impact can be very serious. If a disk fails catastrophically, the extensive volumes of company data held on it cannot be accessed or changed. Malfunction of the system, due to improper control of its operational use, or through accessing data through unauthorised or faulty software, can also result in the corruption of valuable company data. It is essential to have a number of copies of the data for retrieval and data recovery in the event of data loss, corruption or hardware failure.

The 'backing-up' of system data can be considered as two key tasks - one for the archiving of completed work and the other for regular copying of data relating to work in progress on the system. Archiving of finished work will release on-line disk storage space for new work (this will be discussed later within this chapter). The archiving strategy will relate to specific engineering product and project considerations. There is however an immediate priority to set in place back-up procedures for computer aided work which starts to become work in progress on the computer shortly following the introduction of the system.

Different back-up practices can develop within a company and some strategies can have their own hazards over a period of time. For example a simple practice, which may be initially thought of, could consist of simply taking a full system back-up on to interchangeable disk packs, floppy disks or back-up tape cartridges on a weekly basis. This simple approach could cause problems in the event of equipment malfunction, or errors caused by less experienced users.

One of two approaches described in the following paragraphs should be adopted as a company standard procedure. One approach gives a fully covered back-up facility and the other represents the minimum cover for computer held technical data. Both of these approaches assume that equipment includes a tape streamer facility. A tape streamer will minimise the cost of back-up storage media (tape cartridge). If the system is implemented within the context of a local area network, the tape streamer typically would be deployed on the file server node.

The back-up (copying) of computerised work in progress data should take place on a regular basis with a series of cycles covering all ongoing work. The frequency of these cycles is dependent on the amount of data being produced and its value to the company. Two methods are discussed below. In the case of the smaller company which has implemented only a computer aided system of modest scope and complexity (say a single function system such as a simple two dimensional computer aided draughting system), perhaps the second method set out below should be adopted. This ensures that minimum cover is obtained. If however the value of information is deemed by the company to be particularly valuable, the first method to be described should be adopted as a matter of prudence.

It should be noted that the first method described requires extensive administration and a substantial number of back-up tapes.

Full back-up protection

For full back-up and error recovery protection the following set of procedures should be followed. On a four-weekly basis, a full back-up of all data held on the computer aided systems should be taken. This should cover all files and directories, with the exception of those associated with the system's software, such as the operating system and the system's application program software. Copies of these software files should already be held on back-up tapes and/or floppy disks as master disks and tapes away from the working copies of the software. This monthly data back-up should be copied on to a series of tapes on a rolling basis. This means that three tapes (or sets of tapes) are used to rotate on a 'son, father and grandfather' basis. This means that when Tape three (or set three) has been used at the end of month three, the next tape (or set) to be used will be Tape one (or set one) again. This will always provide at least two months of back-up cover.

Also, on a weekly basis over a three week cycle, a back-up of the user project data directories and files should be taken. This will capture all working and project files in the system(s). Again 'son, father and grandfather' copies should be maintained and recycled as for the four-weekly back-ups.

Finally, on a daily basis, between the weekly back-ups, a back-up called an incremental back-up should be taken. This type of back-up only copies the files that have been modified since the previous day's back-up was taken, thus only capturing new or latest versions of files. Assuming a five day working week, a set of tapes (or set of floppy disks) for each day of the first four days of the week are used. These are then recycled on a weekly basis and kept safely away from the system.

This full procedure will require the following tapes;

>Three monthly tapes (or sets);
>Three weekly tapes (or sets);
>Four daily tapes (or sets).

This full and comprehensive formal procedure is designed to ensure that in the worse case, only a day's work would be lost in the event of a major disk failure or data corruption. In addition, for applications which require the users to store their data from time to time (such as computer aided draughting or modelling system) each system user should save their work on the on-line disk system on a frequent basis by invoking the appropriate 'save' command within the application system. Working data should be saved at least every half hour or more frequently if the working session includes extensive editing of the data. For other systems in which the operational data is constantly being updated in real time, the data is automatically being updated within the on-line databases.

As a further protection against corrupting good quality operational data, if say a system user is exploring alternatives within a design scheming session, based on a partly completed drawing or an earlier version of a drawing, it is prudent practice to first take a copy of the previous drawing upon which the experimental design is to be based. Later, the decision can be taken as to whether to allow the new version to be saved and to be re-named as the current version, in favour of the drawing from which it has been derived.

Minimum back-up protection

The following method of taking back-ups should be considered as a basic minimum strategy. Within this approach, a master dump of all user data files and directories is taken on a rolling two week basis on to three tapes or sets of tapes on the basis of 'son, father and grandfather'. These are taken in a similar manner to the monthly back-ups referred to in the first method.

Additionally, on a daily basis, an incremental dump is taken. Here two sub-options exist. The first is to take an incremental dump every day on a rolling basis between the fortnightly back-ups and which backs up the work done since the previous day. For a five day working week this will require nine back-up tapes or sets of tapes (or floppy disks). Each could be holding very little information - only the files which had changed from the previous day. The better option would be to incrementally back-up every day, but to retain the 'datum' of the incremental back-up. This is best explained by an example.

By way of example - on Monday an incremental back-up of that day's work should be taken. Then on Tuesday, a incremental back-up of both Monday's and Tuesday's work should be taken. Then on Wednesday; Monday's, Tuesday's and Wednesday's work is incrementally backed up. This cycle is then rolled forward re-using the three tapes (one day, two day and three day tapes/floppy disks). On Thursday the cycle would start again by taking only a single day's incremental back-up (Thursday's work) on Friday a two day incremental back-up would again be taken and so on.

The computer operating system will allow the incremental back-up to be taken from a nominated date and indeed the incremental daily back-up strategy could be extended beyond a three day cycle if required. It is suggested however that no more than three days' work should be backed up on to a tape/floppy disk set. The system administrator needs some simple method of knowing what day in the cycle he or she is at on any particular day. Either a simple numbered planning chart or a colour code *aide mémoire* normally suffices for this purpose.

This approach will considerably reduce the risk of any problems in continually saving corrupted data on to successive back-ups with its attendant risk of losing good quality, if old, data. It will allow additional resilience within the back-up strategy. This second method variant will require the following tapes:

> Three master back-up tapes (or sets);
> Three incremental back-up tapes (or sets).

The same periodic saving of drawings data etc during the working session on to the system on-line disk should also apply as for the full cover back-up strategy described earlier.

It is important not to allow the cost of buying back-up tapes or tape cartridges to compromise the back-up procedures. These tapes are quite expensive but only a minor expense compared to the system costs and the value of the data which they are protecting.

The management of the company must decide how extensive they wish to make their back-up procedures. Once that decision is made it should be implemented quickly. To continue with inadequate practices for back-up would place the company in some risk of losing valuable data and could result in significant recovery costs. Any loss would place extra work burden on an already limited trained staff resource and would only serve to reduce overall confidence in the technology at a crucial stage of its introduction. The company's system administrator function must be established at an early date and he or she should set up batch files on the system to semi-automate these back-up procedures. These should be supplemented by written customised scripts relating the user and projects directories which are implemented within the company's system in order to encourage the use of these prudent formal procedures.

Extending the training and workload

As the company gains confidence in the use of its computer aided systems and the associated range of administrative procedures become established, workload can be extended and more difficult tasks attempted. Some teething problems will occur with some users of the system and ongoing help from the vendor will be needed in the early weeks and months following the start of operational work on the system. It is wise to plan to supplement the initial training of the operational staff with both refresher training and, in some cases, more advanced training after a period of one to three months. By this time users will have identified areas where additional advice is needed. During these training sessions there is an opportunity for the vendor to identify any bad habits which users have picked up and identify better ways in which their system could be used to accomplish particular operational tasks.

Archiving data

The question of archiving user data files from the computer system was identified earlier within this chapter. From time to time, computer data files which relate to activities completed for some time may cease to become actively used on the on-line disk. The operating system's file handling routines can be used to identify the date and time when a file was last modified, serving as some guide to its activity. However, some of these data files may be referred to as the basis for new work or by other functions within the company than that which originated the file.

From time to time, the system administrator should circulate a list of older data files which have not been changed for some time to all users of the system. Together with this list an instruction should be attached for users to identify data files which can be sent to archive. Other older files may no longer be of any significance to the company and may not even be worth archiving. From the responses, the system administrator can decide on an archiving strategy and arrange a series of archive tapes best suited to the company structure (e.g. by project, project group or the operational department of the originator of the archived file). Having deleted files or transferred them to archive tapes, on-line disk space can be recovered for new data.

Long term storage of archived tapes can be a requirement for a company, particularly where the data relates to defence equipment or where a long term commitment to product or system support must be offered by the company. Care of archive tapes over several years can demand special precautions. For example, under storage conditions, data corruption can occasionally occur on magnetic tapes if they remain unused over many years. One way to avoid these problems is to wind the archive tapes fully forward and then rewind them on a regular basis (e.g. every six months or so). This process helps to minimise the effect of magnetic print-through between adjacent tape surfaces which has been known to cause data corruption. Also, the archived tapes can be regenerated from time to time by loading the data back on to disk, validating the data and then re-transferring the data back to tape.

Monitoring the system

In order to ensure the planned objectives of introducing computer based technology are achieved, it is essential to monitor the use and performance of the installed computer system(s) over time. It is also important to assess and refine the associated procedures and to monitor staff who are directly and indirectly concerned with its use. During the early period, when the company is assessing the need for computer aids and subsequently when technology packages are being acquired, senior and middle technical management invariably take a keen interest in the scope of the technology, its potential functional performance and relationship to the operations of the company.

Like many new tools, interest peaks during the initial exploration stages and can fall off once introduced into a company. After all, technical aspects of computer aids are very likely to spark great interest in technically trained management staff. The potential of the technology can be related to the company in broad technical terms by such management and many will become enthusiasts for greater computerisation. Once the technology has been installed, there is a strong temptation for the broader management of the company to conclude that benefits, which have been much vaunted by the vendor's sales staff, will flow. Subsequently, it is fondly believed, many aspects of the company's technical operations will be improved, if not fully solved.

Frequently, it is assumed that the staff who have researched the technology, put together the case for its deployment within the company, assessed the contending vendors and finally installed the system, need to be left alone to get on with learning how to use the system and then to apply it to the work of the company. Senior management must turn their attention to other matters within the company. They have played their part in the commercial assessment of the technology package and are prepared to trust the operational staff to make best use of it.

It is often some time after the installation of the technology package that senior management may start to realise that all is not well. The normal commercial performance parameters do not show the hoped for improvements in company performance. It must be expected that operational performance may drop during the early months after the implementation of a major computer package. Staff must climb a learning curve with the new system before they become proficient in its day to day use.

Where the new system is replacing an older computer system, or a system based on manual methods, the new system must be run in parallel with the old system for a period of say two months or so as mentioned in Chapter 11. This initial period will place extra demands on the operational staff and functional management within the company. Time must be allowed for the new system to bed down within the company. There will be peak loads during the period when operational data must be transferred to the new system, particularly with materials and production planning and control systems. These effects will be compounded by the operational staff being less familiar with the new system than their previous procedures and methods. There is a need, therefore, to implement major computer systems in planned incremental steps, pausing to consolidate the gains made with each stage on route to full implementation.

Similarly, with systems such as computer aids within the design and drawing office, retrospective work may be used to confirm the best methods of using the new computer aids to meet the needs of the operational staff as discussed in Chapter 11. It is essential to plan retrospective or parallel work. Firm targets for the desired operational results from the new system must be agreed and documented, together with target timescales and performance milestones. The additional staff resources needed during this initial learning time must be estimated and agreed by the operational management. The associated staff training plan must be co-ordinated with the targets set for this initial period.

The learning curve

The learning curve for a new computer aided system will depend upon the scope of the system and the nature of the staff who are using it. For a basic two dimensional draughting system, most operational draughting staff should become proficient in the use of the basic system within a period of four to six weeks following initial training. More advanced applications, including three dimensional work, solid modelling, parametric geometry etc should take a little longer, of the order of three to six weeks. Younger staff, up to the age of about 35, are likely to adapt more rapidly to the technology on average. Older staff can, on occasions, prove to be just as successful with such systems.

Integrated systems, associated with materials and production planning, scheduling, monitoring and control, are likely to be used by a range of different staff within the various associated functions within the company. Many of the operational users are likely to be concerned with a small subset of the modules within the integrated system. Within each of their limited uses of the total system, most of these staff should become proficient within the period of parallel running, typically within two months or so. Similarly, the proficiency of the system administrator should be adequate within about the same period. This however assumes he or she has undertaken full and comprehensive training. Staff who use the functions of the system on an *ad hoc*, occasional, or infrequent basis are likely to take longer to climb their learning curve. They are unlikely to ever become fully proficient in using the full scope of the system.

Close co-operation with the system vendor is needed to ensure that various levels of staff training are completed and that this training is reinforced by on-line help and support facilities from the vendor. Staff who have already received initial training will find areas where they will need further support, guidance and help. The relationship with the vendor during these early months, is therefore crucial to maintaining staff confidence in the system and in reinforcing good working practice with the new system.

Reviewing the operational use of the system

After a computer based system has been installed for a period of about six months or so, a review of the full system operation should be conducted. This review should cover more than the performance of the hardware and software aspects of the system. The range of work and functions of the system being used by the operational staff should also be included. The original planning assumptions made for the operational use of the system should be tested against the actual observations and perceptions. Some of the planning assumptions made for deployment of the technology will relate to other aspects of the company operation. For example, plans to rationalise the company product range or to build in more modularity into products and systems marketed by the company may have been made - have these changes actually taken place?

In order to conduct such a review it may be appropriate to call in specialist skills either from the vendor or from the independent consultancy organisation which may have helped in the original assessment of the technology needs of the company. Much will depend on the level of appropriate technology skill which, hopefully, has grown within the company.

With regard to the hardware aspects of the system, overall performance and reliability should be assessed. Any equipment failures should be logged over time and reported to the vendor. These records should be reviewed and any recurring problems identified and highlighted for detailed discussion with the vendor.

Similarly, utilisation of the workstations and other terminals should be logged, either by automatic facilities provided within the operating system or by some simple form of manual record of usage. The usage figures should indicate where the access resources to the system have been both under and over provided for in the initial planning. Such basic usage figures need to be considered in light of the output and performance of staff who normally use them. For example, a terminal or workstation may be loaded more highly, not because the work throughput is higher than originally planned for, but because one or more of the staff which normally use it is making many more mistakes or the response times are appalling. Such users may be spending a lot of time correcting their work at the terminal. Or they may be attempting to do more difficult work than their level of training will support.

The disk utilisation of the system should be monitored and reviewed against the projected utilisation for the existing stage of system implementation. For computer applications which contain a high degree of graphical data, (e.g. draughting or topological or analytical modelling applications), the data file sizes which are generated by different users of the system should be monitored and compared. For operational tasks which are of similar complexity, but which have been performed by different technical staff, it may be found that data volumes are markedly different between these user staff. This may lead to a conclusion that certain staff are using the system more efficiently than others. The methods used by one user can then be compared with others and operating practices refined and optimised within the company.

Monitoring response times

The response times of the system and the consistency of these as the number of concurrent users varies, are key parameters to monitor. These will be of particular importance for systems where the computer power and/or the communication bandwidth between the computer and its users is shared amongst a number of users. If problems of response time are detected at this early stage, it bodes ill for the future. Data volumes are likely to grow and the demand on the system may well increase from the effect of releasing suppressed demand for the facilities of the system. The suppressed demand effect is often a factor when the system is used more and understood within the company. Both of these future influences are likely to increase the load on the system and will further adversely affect any early problems of inadequate response times.

It is essential that if system response time problems are detected at this early stage the facts should be recorded and communicated quickly to the vendor whilst the system is still under warranty. Hopefully, the contract of supply has been agreed as previously set out. In part, it should be based around functional performance guarantees on such key parameters as response time. If so, the company will be on stronger ground in seeking a satisfactory remedy from the system vendor. Unsatisfactory response time is one of the most common causes for computer systems to fail to live up to a company's expectation. Vendors have had a reputation of under resourcing the computational power of technical computer aids in their initial supply of equipment. If problems are suspected, these must be nipped in the bud. The unresolved problems are likely to become much worse in the future if they are not resolved at an early stage.

One alarming characteristic of the deterioration in response time, in systems where the computer power is not widely distributed to the system users, concerns the manner of response time degradation with increasing computational load. System response times can deteriorate quite rapidly with increasing computational load when this exceeds a maximum acceptable load. The effect is highly non-linear. Suddenly, the addition of one further user of an application which demands a heavy computational or disk access load, can render the entire system virtually unusable to other users sharing the computation resource. Hence the strong preference for a hardware system architecture based on distributed processing. In these configurations, the heavy users of processing power are serviced by dedicated computer processors.

Reviewing data security procedures

The operation of the data security back-up routines needs to be reviewed and audited. Bad habits in this area can cause extremely serious problems in the event of unforeseen problems at some later time. Deficiencies in the operation of these routines should be rectified and, where appropriate, additional staff training should be undertaken. The labelling and storage of off-line data needs to be monitored and any deficiencies which are identified must be corrected. Standard operating procedures for handling data security and error recovery strategies need to be checked for clarity and lack of ambiguity. The nominated staff charged with performing back-up routines should be checked to see that they are fully familiar with what is needed in this essential activity.

Application software performance monitoring

Just as all malfunctions of the hardware should be logged and reported to the support organisation, so too with the application software. Most vendors will supply the company with standard report forms upon which suspected operational errors within the application software must be reported back to the vendor. Clear recording of software errors should be logged by the operational staff and co-ordinated via the system administrator, back to the vendor. For the larger integrated systems, many vendors will provide remote access from their support site into the company installation over a modem link which operates over the public switched telecommunication service network. This allows the vendor's support staff to gain access to the system within the company and to explore suspect operation of the application software. Furthermore, possible sources of corrupted data produced in conjunction with the system can be assessed. Such remote modem support facilities can save time in resolving problems and help in eliminating unnecessary visits to the company site due to operator error or other causes.

It is frequently the case that most users of computer aided systems initially use a relatively small subset of the total range of commands which are supported by an application software module. Many users become familiar with a small range of commands and functions with which they quickly gain confidence. These are typically the more basic functions. Often there are more refined and elegant functions and methods which are provided by the system. These underlie the system and can offer the basis for achieving the maximum effectiveness of the system. Their use may demand a more in-depth understanding of the system. It is often the case that the operational result obtained by these more elegant functions can be achieved by combinations of more simple and basic functions. However the latter approach may be much less efficient and more time consuming.

It is important to pick up the cases where users are still intimidated by the system and do not use the more effective commands and combinations of procedures which are available. Once again a mature user of the system or the vendor can help to identify these cases. Timely retraining can help to eliminate poor operational use and practice with the system. Users will learn from each other, to some degree, how the application software can be used more efficiently. However, a number of users are likely to be less outgoing. They may fail to develop their skills with the system in the absence of an external stimulus aimed at optimising their methodology.

Many of the more successful computer aided systems have attracted strong user groups. Participation with the user group should be encouraged. Not only can the group provide an extremely helpful source of information on the practical operation of a system but it can serve as a co-ordinated pressure group on the vendor. The user group may serve to identify the sources of helpful additions and extensions to standard application software packages. These include libraries of information related to particular engineering disciplines, less standard interfacing routines between modules and customised command macro routines.

The user group can therefore provide a method of monitoring the performance of the company's user staff with others working with the same computer aids. It can help to eliminate simple operational mistakes and bad habits in using the technology. The group may serve to identify future development paths for the technology within the company, based on the previous experience of other companies. The user group discussions and its collective observations help the vendor formulate their own development plans and to monitor how their computer aids are succeeding within the market place.

Technology champions and antagonists

Within a company which has recently embarked on upgrading its technology base, the natural tendency amongst the management is to encourage the enthusiasts for such technology to press ahead with both deeper and wider use of the computer aids. Technology 'champions' naturally emerge within the company. These are found within the management of the company and within the operational staff. Often these tend to be drawn from the younger end of the age distribution, although converts amongst the older staff are not infrequent. Perhaps the initial enthusiasm has arisen because of visits to exhibitions, conversations with others on the golf course or in the pub. Danger can result when the technology champion is only superficially aware of the strengths and weaknesses of the technology and has not been a part of the company's team which has rigorously assessed the technology requirement of the company.

The expectations of these less than fully informed enthusiasts can be quite unrealistic in the early period after installing a system. Their initial enthusiasm can quickly turn to antagonism to either the system or the staff using it, if their expectations are not realised. It is so often said that the performance of any system can only be as good as the information provided to it. It also depends on the timeliness of this information and commitment of the staff to make the system work to the benefit of the company. Antagonists to the system, located in sensitive areas associated with the operation of the system, can quickly become dangerous loose cannons within the company. The disillusioned technology enthusiast can be more disruptive within a company than the person who has been sceptical about the use of the technology aids from the outset.

Middle management attitudes

Section leaders and middle level managers, particularly within design and drawing offices and within such functions as process planning, have traditionally discussed the work of their subordinates whilst observing the work as it evolves at the subordinate's desk. Drawings and other documents manually prepared were clearly visible at the scale in which they were being drawn. The section leader or manager had no problem in relating to the information being prepared and the methods used to prepare it. They could look over the shoulder of their staff or walk around at the end of the day to assess the progress which had been made. They knew from their own experience how long a task should take and could recognise when the subordinate was having problems or when additional guidance should be offered.

With the advent of computer aids within a company, detailed work progresses within the computer and is frequently only visible on the screen when the subordinate is working on it - and then only one screen at a time. Thus, the evolving information is less immediately visible to the section leader on a casual basis of inspection. Section leaders tend to be drawn from the older staff within the company. Generally, they will have extensive experience of the company, its products, methods and standards. Within some companies, senior operational staff have tended to take a rather hands off approach towards new technology packages which have been introduced into the company. There is a temptation for the section leader or middle manager to take the view that; since they spend a relatively small amount of time in actually committing design information in the form of geometry lines and text on paper, either at the drawing board or the desk, there is little point in learning to use the new system. They consider that their subordinates must become proficient in the use of the new systems but they themselves need only know what, in general terms can be done with the new computer aid.

If this attitude prevails, a new barrier can be built between the operational staff and their immediate superiors. The confidence of the section leader in day to day interaction with staff who use computer aids can be subtly eroded. Gradually, section leaders can become somewhat intimidated by the technology if they fail to develop a basic competence in its use. At least they should be able to efficiently interrogate the information which is being developed by their subordinate staff. Recalling and manipulating different views of drawings, exploring of the relationships between information held on the computer and gaining access to information concerning the usage of their staff's equipment, must be considered the minimum skill of the section leader or middle manager if they are to maintain appropriate control and supervision. It is not really satisfactory for the senior operational staff to rely solely on accessing the work of subordinates only when finished sets of documents are produced as hard copy output from the computer system. They must be able to interact with staff and monitor work in progress on the computer. This requires training to a basic level of competence in the use of the computer aids which staff are using. It is also important that this training is reinforced with day to day usage of the system.

The section leader or office manager is unlikely to become as proficient in the full use of the system as staff who work with the system regularly. As with any other tools or equipment used within the company, it is part of the responsibilities of operational management to ensure that best use is made of them. They must acquire sufficient knowledge to discharge this responsibility and the senior management of the company must ensure that middle management receive appropriate training and exposure to the system. As time passes, and section leaders and middle managers increasingly become drawn from staff who themselves are familiar with computer aids and their associated techniques and methods, this problem will be reduced. In the mean time the management of the company must recognise the problem and plan to remove it or at least minimise its effect by prudent training and counselling of the older middle management staff.

Monitoring the administration of the system

Unlike large companies, which tend to manage their computer systems within a specialist service section of the company, small and medium sized companies cannot justify such a specialist department. The medium sized companies may have developed a small commercial data processing section, typically within the financial and accounting function. In this section staff may be found with some system design and software programming skills. Within the technical functions of the company, staff are increasingly found with some technical software development skills. The advent of significant computer aids for the technical functions of the company can pose new questions. Should the new systems be controlled and monitored by the commercial data processing section? Or should they be left entirely to the engineering and technical staff within the technical functions of the company?

Skills associated with commercial computer based data processing systems typically have only a superficial relevance to those needed for rigorous control and further development of computer aids associated with the engineering functions of the company. There are general areas of overlap, such as database management, back-up procedures and hardware assessment. Computer orientated staff who have gained their experience in commercial data processing and accounting functions tend to have difficulty in communicating in a satisfactory manner with the technical and engineering staff using technical computing aids. On balance, the best approach is to monitor the performance of the technical computer aided systems within the operational functions which are served by them, co-ordinated by a technical system administrator. There are areas of application overlap between technical computer systems and those more naturally concerned with accounts and the financial functions. This is particularly true when these systems merge into the topics of sales and purchase order processing and stock control systems. Clear responsibilities must be determined and the computer related skills should be deployed in some form of matrix structure in order to improve the performance of existing systems and to plan for future enhancements and developments.

Monitoring the supplier of the system

Not only must the performance of the technology packages be monitored within the operations of the company, but the performance of the suppliers who supplied them should be recorded and analysed. As was identified earlier, application software within technical computer aids invariably contain a very large number of lines of software. Even modestly sized applications will contain many tens of thousands of lines of software code and some will contain hundreds of thousands of lines of code. The number of different sequences taken through the software during its day to day operation within a company, will also be large. It is a fact of life that some errors will exist within the software. The developers of the software will have tested the software thoroughly before releasing it to the market and will have eliminated the vast majority of errors which are initially introduced in writing or amending the software. However, a small number of errors will remain, resulting from particular combinations of sequences of operation or data.

Certainly, no errors should remain which cause the system to fail in a manner which renders the user unable to continue with normal operations of the system. Such fatal errors in the software would be a cause for rejection of the software and may have been identified during the benchmark testing of the system. The more likely situation would be that occasionally, particular functions, in some circumstances, will fail to work as set out in the operations manual. Often it is possible to achieve the same result as that which is being sought, by using an alternative sequence of command functions. Also, as with any other electronic equipment, hardware units within the system can fail.

Each hardware or software problem which is encountered by the company should be reported to the vendor, initially over the telephone. If on-line modem support is provided, the vendor will be able to explore the problem from the support site. If this level of support does not resolve the problem, it should be reported formally in writing to the vendor. This will normally be done using a standard reporting document provided by the vendor. The report should identify the precise perceived nature of the problem and the context within which it arose. On completion it should be dated and signed. If appropriate, sample output documents and files which relate to the problem should be attached to the trouble report. A copy of the fault report and associated documentation should be sent to the vendor and a second copy filed by the company system administrator.

Under the terms of the hardware and software support contracts with the vendor, reported faults will normally be categorised typically as being of one of three categories. The first is the so-called 'fatal' category, which results in the system needing to be restarted with an attendant loss of all information since the last 'save' or back-up sequence was performed. The second type of fault is said to be 'urgent', - for example a particular command function does not work in the manner described. The third category can be considered as non-urgent, - a command does not perform quite as expected in certain circumstances, but alternative methods exist to meet its objective, or it may be that the function would be preferred if it worked in a slightly different manner.

On receipt of a fault report from the company, the vendor should be expected to identify the prime source of the problem in terms of whether it is hardware or software based. Hardware problems should then be remedied in accordance with the timescales set out in the maintenance contract. These hardware problems will normally require a visit by the vendor to resolve the problem. Software problems will be mutually classified between the company and the vendor as to which of the three types the problem comes under. Clearly the 'fatal' software errors are extremely serious and should not exist in released professional grade software. If they do occur in a software update release, the vendor should be expected to make best endeavours to resolve the problem as a matter of great urgency. Either a solution to the problem in the form of a software patch, or a different but data compatible version of the software, should be rapidly provided.

The vendor should respond to the urgent category of software error within a time period specified in the software support contract. Either a 'work around' or a software patch should be sought within this period. Such problems should be fully resolved within the structure of the next release of the application software. The 'non-urgent' problem will normally be resolved only in the next release of the software. If it is more in the nature of a 'wish' by the company, rather than a clear fault, the vendor will consider it in the light of their own plans for the future development of the software.

The system administrator should monitor how well the vendor responds to the various types and classifications of problem encountered. An analysis of vendor responses should be performed and documented on a regular basis, say half-yearly or annually. If it is felt that the vendor's support performance falls short of the commitments set out in the support contracts, the vendor should be appraised of the facts and figures of the shortcomings. It can also be useful to compare the support performance of the vendor with other users of their system, through enquiry and discussion within the user group. If support performance continues to be a problem, the matter should be taken up with the vendor at a senior level within their organisation. This can be done either directly from the company or in consort with other members of the user group if appropriate. Failure by the vendor to adequately support their product can result in serious limitations to the likely benefits which can be realised by using the system.

Decisions on expanding the use of computer aids

Over the longer term, with proper monitoring of: the computer aided systems, the competence of the trained staff which operate them, the skills of the middle management who review these staff and the support performance of the vendor, the management of the company can form a view as to when the maximum benefits from the system have been achieved. These benefits should be expressed in terms of, and compared with, targets set out in the original technical and commercial justification for the technology. These will normally relate to the various technical and commercial functions within the company. The success or failure of the technology package to meet the original objectives, as perceived by the management of these functions, should be documented in a formal review report. This document should then be discussed with both middle management and key operational staff.

If the outcome of the investment has been generally satisfactory, consideration can be given to expanding the scope and/or the depth of the application of the technology. Technology upgrade paths need to be identified amongst the functional management of the company. If the original analysis has been thorough, these future developments will have been considered in outline at the outset of exploring the initial technology package. An implementation plan for these upgrade paths needs to be documented. It should also be provisionally resourced, prior to further discussion with the original vendors of the system(s). Hardware upgrades and extensions should also be explored and discussed with the original vendors of the initial technology package.

As well as identifying additional standard proprietary application software modules, from which the company could derive further benefits, any company specific application modules should be considered. Often, these will be based on standard software modules and will be extensions to them or variations which meet specific needs of the company. Alternatively, perhaps in the case of specialist analysis requirements, fully bespoke software modules will be needed to be commissioned or written in-house. In either case, there are no short cuts or substitutes for fully defining the functionality of the application module which is needed together with the ranges of data over which it is needed to operate.

Possible reasons for poor results

Not infrequently, when a review is conducted, it is perceived that the performance of the technology falls below original expectations. If the result has generally not been satisfactory across a range of company functions, the reasons should be sought. There are a number of possible reasons why the introduction of computer technology fails to meet the much vaunted objectives. The following list of explanations have been found to be underlying causes when such reviews have identified shortcomings:

— Poor system response times when fully operational, due to inadequate hardware resourcing of the system;
— Inadequately trained staff who are under utilising the functionality of the system;
— Badly selected application software;
— Failure to implement sufficient data within the scope of the system to produce optimum operational results;
— Inability to link application software modules together;
— Poor system support from the vendor;
— Poor user documentation;
— Poor management and administration of the system;
— Poor management commitment to resourcing the system;

The above list is quite revealing. There are still a significant number of cases where a company needs computer aids like a hole in its corporate head! This is largely because of the nature and mixture of their work, or because they do not have the basis of the disciplines within the company upon which a computer based approach can build. In this latter case it is not possible to enhance the communications and controls within the company. Much more frequently however, the root of the problems is to be found more with the people than the limits of the technology in which the company has invested. Sometimes two companies, or even two sites of the same company, engaged in similar work and which have installed identical (or very similar) technology packages, perform quite differently in terms of the degree of success achieved. When the causes are explored and analysed, the basic reasons behind the different performance are concerned with the management and operational staff, their skills, perceptions, organisation and commitments to the technology.

Conclusion

The introduction of computer based technology cannot be considered simply as the introduction of a new tool within the company. The disciplines needed to make such systems succeed have a wide impact across the company. These disciplines need not eliminate flexibility if the computer based system has been chosen appropriately. Indeed, one of the major benefits of a properly implemented computer based system is the ability to explore alternative implementation and control strategies and to identify the likely consequences of these options. The key point of the computer assisted approach, is that exploration of alternative approaches to control is undertaken within a structured environment.

As has been indicated earlier, a computer based system cannot in itself impose structured methods and disciplines within a company. The system can only respond to the prime company information fed to it. This information must be generated in an orderly manner and made to flow in a structured way throughout the company. A computer approach can reinforce the disciplines which have been introduced within the company, but it can never be a substitute for them.

The introduction of such technology demands that the methods, customs and practices within the company need to be refined to ensure that adequate information is available for the system and that it flows across inter-departmental boundaries. Its use may suggest rationalisation of design practices and procedures within the company and can facilitate these changes. The precise nature of the changes will of course remain under the control of the operational management of the company. It is important that each of the operational management team acquires a clear and detailed understanding of the potential for gain from the introduction of the technology and does not abdicate his or her responsibility to others within the company.

The clear implication is that there must be a corporate commitment to the introduction of the technology. An informed consensus must be formed across the functions of the company as to how to implement new methods and reinforce existing disciplines and procedures. A corporate strategy for the use of such technology packages must be formulated. This must have within it a plan for achieving full and successful conversion to the new systems. The short and medium term plans for developing the technology within the company must be clearly set out and progressed from the highest levels of management within the company.

13

The problem areas

Despite the rapid development of these technologies over the last 25 years problem areas remain. As a community, engineers and technical staff are still learning how to make best use of their systems for better engineering practice. Likewise, developers and vendors of systems are still feeling their way to improving the functionality of their systems and making the interface between the technical practitioners natural and easy to use. Within this chapter some of the problem areas associated with the technology will be discussed.

The sources of these potential difficulties are various. They are associated with the following general aspects of exploring and adopting the technology:

- The supply market;
- Converting to using the technology;
- Components within the technology;
- Interfacing users with systems;
- Interfacing systems with systems;
- Perceptions relating to the technology.

The supply market

Perhaps the first area of difficulty which any firm finds when it starts to explore the technology is the wide range of companies offering computer aided solutions to an ever increasing range of engineering related problems. The person within a company charged with exploring the needs of the company, who is approaching the technology for the first time, can rapidly become confused. Should concern be centred more on hardware or software? What does all the jargon mean? Why do products on offer range in price by over a factor of ten for a system which appears to offer similar capability? After all, the promotional material for, say, different so-called CAD products each shows complex drawings and diagrams which, it is claimed, have been produced by the system on offer.

The trade press associated with the technology is of limited help. The visual appeal of computer graphics in glorious colour forms a powerful stimulus to the advertising community. The trade advertising for computer aids can often be just as overstated as any of the excesses to be found in fashion magazines or children's magazines. Trade exhibitions can be a source of information on the technology, but all too often the technology is shown in the context of pre-prepared canned demonstrations. These are of limited value for gaining any in-depth view of the potential of the system with respect to the needs of the company. Even where the system is being demonstrated in a more free format, with interaction between the demonstrator and the public, it is often being demonstrated in an artificial environment. The operational conditions within a company are likely to be quite different. The computer hardware resourcing of the system will be optimised for the purposes of demonstrating the system at an exhibition. The size of operational data files are typically quite small for demonstration purposes. Such demonstrations are as much a display of the demonstrator's competence as they are of the system. A well rehearsed demonstrator of a computer aid will be very good at showing off the versatility of the system and its strengths whilst subtly hiding its limitations and weaknesses. It would be a brave person indeed who recommended a system to their company, based largely on press advertisements or what was seen at an exhibition.

The volatility of the supply market is still a notable feature. This is despite the periodic rationalisations of the supply companies which have taken place, particularly during times of recession. The investment in such technology needs to be considered as a key element in the overall development strategy for the company. It will often be a staged investment over a number of years. There are attendant problems of ensuring that the chosen supplier is going to be able to offer upgrade paths to support potential later stages of system implementation. If the supply company ceases to trade or their system merges with that of another supplier, there are potential pitfalls of system obsolescence in the future.

This problem relates more to the application software content of the system rather than to its hardware. It is a pretty safe assumption that the hardware content of the system will become, to a degree, obsolescent within a period of less than five years from the purchase of the system - such has been the pace of computer hardware development. This in itself need not be a cause for great concern provided that the application software is designed to run under a widely used and mature operating system environment. The hardware content can be upgraded or even changed altogether whilst maintaining compatibility with the operating system and hence the application software.

Whilst there can never be any firm assurance that a supplier has a long term potential for commercial survival, the normal commercial assessment of a supplier can give some feel for their market share, profitability and growth. The very prudent advisor will normally recommend the tried and tested larger supplier. After all nobody ever got fired for recommending IBM! However, not too long ago the same could have been said in a different market sector of Pan Am.

There is no substitute for assessing the various vendors' offerings against the researched and documented current and longer term needs of the company, as has been described earlier within this book. Having identified the best technical match with the company's needs at an acceptable price, normal commercial vetting of the supply company is all that reasonably can be done.

Converting to using the technology

Having eventually made the decision on what system to purchase, the company is faced with the problems of converting operations to take advantage of the new technology package. A number of the issues to be considered in setting up the functions of the company were discussed in the previous chapter. Care must be taken to make the operational units within the company, and where appropriate, the representatives of organised labour, aware of what target benefits are being sought by introducing the technology.

Problems can arise in the transfer to computer aided methods when insufficient planning has been undertaken. A full operational commitment to the new computer based system and a period of retrospective work or parallel running with both the new method and the old, can minimise early operational difficulties. Problems arise from a lack of operational experience or less than a full understanding of the procedures which must be followed. A lack of data which the system demands can spell disaster. Not only can the results from the new system be unsatisfactory but the overall credibility within the company can be seriously damaged.

Similar problems can arise where staff have either been inadequately trained in the use of the new system, or where training has taken place too late or too early with respect to the installation and use of the computer based system. For the more comprehensive computer aided systems, training ideally should be undertaken in a progressive manner over the period of planned implementation of the system modules. If problems are to be avoided, additional formal training by the vendor, or other professional training organisation, should be sought when the need is identified, despite any additional cost.

Staff performance monitoring has been discussed previously and it is essential to identify those staff which fail to ascend their learning curve at an acceptable rate, and to remedy the causes of the problems they are experiencing. Similarly, bad practice in using the system should be identified by the appropriate section leader or middle manager, and these practices must be corrected before they become endemic. To fail to identify these problems can result in limiting the overall performance of both the staff and the system, to the detriment of staff satisfaction and functional performance.

Problems of organisation and administration of the new system can occur if appropriate staff have not been identified and given the necessary additional training for these purposes in good time. The importance of adequate back-up procedures and error recovery routines has been identified in the previous chapter. Failure to implement these procedures can result in major problems in the event of the loss of data.

Many companies will have a backlog of company data which it would be useful to load into the new computer system. This data may be held in a magnetic form, having been used within an earlier computer system, or it may only exist as a series of manually prepared documents. If the data is in magnetic form, the problem of transferring it to the new system may be partially solved by the use of translation software, provided by the new system vendor as part of the contract of supply. If it is only in the form of printed documents, there is the problem of inputting the data using the normal manual data input facilities of the new system. Whilst optical scanners are available, it has to be remembered that such devices only generate a statistised image of the document in the form of an array of picture elements. The normal ubiquitous scanner device cannot as yet interpret the data into the character format required for holding the company data in a form with which the application software can interact.

Thus, it is necessary to be highly selective as to what, if any, of the prior existing company data is transferred into the new system. Clearly, in the areas of computer aided design, modelling and draughting, frequently used libraries of graphical symbols, (which cannot be acquired as already existing standard libraries with the new system) would be strong candidates for early input into the system. Beyond this, it is doubtful if it is worthwhile inputting much of the other company drawings.

Textual information will need to be input manually in order to set up the master files and databases which are associated with the application modules. Trained resources must be found to initialise the system with the appropriate company data and to validate it, prior to running live with the system. All of which can pose quite a problem.

Components within the technology

Problems associated with the hardware and software components of such systems can occur. Hence the extreme importance of adequate back-up procedures.

Hardware problems

Hardware problems can result from equipment failure or the hardware becoming under resourced for the application. Most of the common sources of hardware problems are likely to be associated with electromechanical devices, such as magnetic disk or tape data storage units, printers, plotters, keyboards and cables. Many of these devices need to operate within high tolerances and can fail unexpectedly, causing severe problems until they are rectified. For example, if a hard disk unit fails catastrophically all of the software and data held on that disk will be lost. This can result in the complete shut down of the system. After the disk unit has been exchanged, the new disk will be empty and it will be necessary to recover all of the software and data from the back-up copies which have been held. This process can be a major operational problem but nothing like the problem which would result if the only available back-up copies are significantly out of date!

Similarly, if the tape streamer unit associated with the system fails, the capability of taking regular back-up copies is likely to be severely impaired. Attendant future disaster can ensue if the problem is ignored for only a short period of time. Such hardware failures need to be put into the context of both their immediate and longer term impact on the commercially safe operation of the system. Remedial actions need to be taken accordingly, normally by invoking the hardware support contract arrangements which need to be reviewed and renewed on an annual basis.

The long term stability and integrity of company archival data held on magnetic tape archives is sometimes seen as a problem. Tape manufacturers will not offer long term guarantees on data storage beyond about six months or so. However, in practice problems rarely arise with this archival data if the precautions suggested in the previous chapter are followed (i.e. the tapes should be wound to their end and rewound from time to time). Routine and regular tape streamer unit maintenance will ensure that tape heads remain within their correct alignment tolerances so that the data can be read back from the tape accurately.

Hardware problems less central to the operation of the system, such as failures of individual terminals, or other duplicated peripheral devices, can be overcome in the short term by sharing resources until the hardware device is repaired. This may result in degrading the total operational performance but is likely to be less severe. Much will depend on the overall size of the system. In any event, equipment failures should be rectified as quickly as possible in order to maintain the system in an optimum condition.

Software problems

Major operating system or application software problems should be rare, provided that a mature system has been purchased. As has been identified, particular individual problems may occur within the application software but the number of these should be small. Often the vendor provides a 'patch' to the software in order to temporarily solve the problem pending a more permanent solution within the next formal release of the software. Where such a solution is not possible or practical, the problems which the software error causes can normally be overcome by alternative methods available within the scope of the application software.

Updating software

From time to time the vendor of the system will issue new releases of application software, together with updates to the software documentation - typically about every six to twelve months for a mature and stable software product. The new release often includes additional functionality within the application software. It will normally provide permanent solutions to the problems which have been identified in the version it replaces. Such releases may also be associated with an upgrade to the version of the operating system software.

It is a normal requirement, within the software support contract, that a user company undertakes to update their implementation of both the operating system and the application software accordingly. This protects the vendor from having to support all the previous versions of the software over a protracted period of time. Typically, the vendor will support only the current version of the software and the one immediately preceding it under the terms of the software support contract. It is normal practice for the vendor to provide a data conversion routine where required, to ensure that data produced by the previous version can still be used with the new version. This is required if changes in data format have occurred with the new version. This practice can result in a source of inconvenience for the user company, since it will be necessary to periodically upgrade the version of the system and to validate that the new version remains compatible with the existing company data.

Needless to say, the new version is likely to contain a small number of new errors. The user company thus exchanges a set of known problems within the old version, for a set of new and unknown errors in the new version. Under normal circumstances the company will however potentially benefit from the improvements within the new version of the application software.

On some occasions, the introduction of a combination of a new release of operating system with an existing or new release of application software can give rise to particular problems. It is not unknown for such a combination to result in a significant deterioration in system response times. This can be due to a major increase in the size of a new version of an operating system (in order to support more functions) and other factors. The effect can reduce the available random access memory within which the application software is required to service the users, and hence the risk of adversely changing the speed of response to each user. Such problems are not entirely rare and can pose a major difficulty for the user company. They may, for example, involve the company having to upgrade the hardware or system, with an attendant additional cost, merely to retain the prior performance of the application software. This can be particularly annoying.

Interfacing users with systems

Despite the major developments which have taken place within hardware and application software, the operational interface between the user of a computer aid and the system's application software, in which the functions are enshrined, still demands some adjustment by the user. Much has been claimed for the use of multiple levels of pop-up and pull-down on-screen menus which were developed over 20 years ago. Their use, with advanced screen windowing techniques, is now ubiquitous, in applications ranging from software aimed at teaching children, to advanced computer aids for business and engineering.

It was believed that the use of on-screen windows and dynamic menus, from which the user can select command functions by means of pointing a mouse driven cursor, removed the need for the user to use a keyboard to any great extent. Indeed, many users find the use of such facilities, or a similar technique of pointing at menus held on graphics tablets helpful. They allow the use of graphical icons, (or pictures) to associate commands with the purpose of the command function within the user's mind. For example a picture of a pencil might indicate the 'draw' command. Certainly, they offer an advance over the rather verbose command sequences which previously had to be input at the keyboard. However the interface between the machine and the engineering user still remains a potential problem area. The acceptability of the man-to-machine interface is a very personal matter. Different users prefer different techniques.

Many mature users tend to move away from using on-screen menus, in preference to inputting cryptic commands from the keyboard. This process seems quite mysterious to the less mature user and can act as a barrier to learning the system if this is all that is seen of the use of the system. One of the problems with pop-up or pull-down on-screen menus is the user often has to go through more than one level of menu in order to perform a particular command function. The commands are arranged, by necessity, into a hierarchy within the menu structure. What is more, the user has to become familiar with the route or sequence through the menus in order to arrive at the command desired. This process can involve a somewhat protracted sequence of cursor selections, with a consequent sequence of mouse movements. The power of each function within the system depends both on the functionality of the system and the structure of the user menus which have been implemented - initially by the vendor but probably added to, over time, by the company's system administrator.

The time taken to execute these sequences of menu selections can become tedious to the frequent and mature user, each cursor selection causing only a modest amount of the work to be done. The alternative input of a small number of keystrokes, directly from the keyboard, albeit in the form of cryptic commands, without recourse to using screen or tablet menus, can often accomplish the desired functional command more quickly and accurately. In the case of computer aided draughting or modelling, the interactions will often require dimensional, tolerance, or textual information to be input into the system. This itself involves frequent input from the keyboard in any case. In other applications for which the inputs are more textual in nature, the need for frequent use of the keyboard will also be necessary. In any event, the younger users of the technology have grown up with keyboards and are not intimidated by them.

In particular, describing geometry to a computer, using cryptic language commands from the keyboard, or by frequent selections from screen or tablet menus, interspersed with entering numerical values from the keyboard, is a necessary compromise. These methods are still seen by more than a few as unnatural compared with picking up a pencil and drawing a line of measured length along a ruler. Similarly, manipulating complex doubly curved surfaces by 'pushing and pulling' control vectors on a mesh of polygons, is probably more unnatural to many users and can be the source of problems.

The building of full three dimensional models, using the available input methods, remains an acquired art. Whilst many staff adjust to using the available commands some, otherwise competent staff find it a problem to make the necessary adjustments. In all such modelling and draughting systems, users tend to limit their use of commands to those with which they have become familiar. This can often limit their overall effectiveness when using systems which, optionally, can be used more elegantly and effectively with a wider range of commands available within them.

The problems associated with the interface between the user and system remain a focus for further research and development amongst system developers. These are difficult areas of the technology to optimise. The reactions of users to different forms and strategies for user interfacing varies, depending, to some extent, on their background experience. The aptitudes which the user has is also a factor of some influence. Since these responses are so subjective, it beholds the vendors to offer a range of optional, and user selectable interface techniques. Some vendors do and some do not. This may be a consideration when selecting or customising a system for use within a company.

Interfacing systems with systems

If the problems of achieving satisfactory communication between the user and the system has been a problem, the problems of transferring information between systems which have been supplied by different vendors can be, on occasions, even more so. As the technology spreads throughout the major industrial sectors, the need arises to extract data from the computer system of one company and read it into a different system in use at another company. Within many major industrial sectors such as the automotive, aerospace and chemical process industries, there exists a hierarchical structure of companies. The major companies in these sectors make extensive use of specialist sub-contract suppliers. These companies are typically engaged in the design, fabrication, manufacture, assembly and testing of components, products and sub-assemblies used within the products and systems of the major companies. As more of these engineering sectors use computer aided methods, there is pressure to transfer computer compatible information both to and from the sub-contracting companies. Many of these sub-contractors are small and medium sized specialist companies.

The nature of the data which needs to be transferred will be a mixture of textual, graphical and various types of computer model data. The model data will typically define some form of topological, logical or functional model of products, sub-assemblies and systems. Earlier in this book the advantages of transferring data between companies in computer form were identified. The key advantage is that each of the companies can work on exactly compatible product data. Data interpretation errors due to human perceptions can be minimised and information can be quickly made available to the technical staff in a form in which they can perform their tasks upon it.

The problem of information transfer is one of data compatibility between different vendor's systems. Each system has its own data structures within its databases and file structures. It also has its own forms of representation for the various types of entity which it handles. These range from the record formats for simple alpha-numeric information such as stock records, customer details through graphical entities such as splined curves and dimension lines and to full solid model entities such as extruded solids and prisms. Indeed, the format and scope of these various types of entity models often underlies why one system is superior to another in its ability to perform some application functions. The databases and file structures which hold information for similar operational applications, used by any two different systems, will be invariably different. Thus the data extracted from one system is incompatible with another. Indeed, data produced by an older version of one vendor's system may not be directly compatible with the data required by the current version of the same vendor's system. In this case however, the vendor will normally offer a data translation routine to achieve compatibility.

Non-compatibility of data between different systems has been a problem in a number of industries. Two approaches are employed to over-come it.

Dedicated link

The first is to obtain a specific translator which converts specific classes of data, such as drawing data, between one nominated version of a vendor's system and a nominated version of a second vendor's system. Within some industrial sectors, where data transfer is a major consideration, a small range of vendors' systems have tended to be adopted within the sector. Data can then be transferred between this small range with dedicated translation routines.

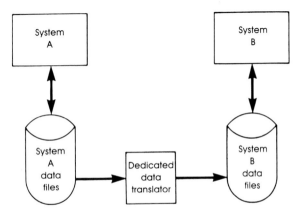

Data transfer through a dedicated link.

Neutral format

The second approach is a two stage process which uses a 'neutral' format to aid the transfer. This method has been much researched and developed for transferring graphical data, found in design and draughting systems.

In the first phase, the data from one system is converted to a standardised neutral data format, using a translator module provided by the vendor. In the second phase the neutral format data is interpreted by a translator, provided by the vendor of the second system, into the internal formats required by the second system. The neutral format must be capable of handling a range of options for modelling the same type of entity supported by a range of different vendors' systems. For example a circle can be defined and modelled in a number of different ways (e.g. centre and radius, passing through three points). The range of these options needs to be widely agreed both amongst the vendor and user community. Hence efforts have been made to agree national, and now international standards for the neutral format. These standards were first attempted in the mid-to late 1980s (e.g. IGES), for simple two dimensional drawing data exchange. More recently these formats have been extended to cover the more complex solid models and doubly curved surfaces together with their associated text. Other forms of neutral interfaces have been developed which are related to particular engineering applications, for example the transfer of data for electronic design including integrated circuit and printed circuit board design data (EDIF).

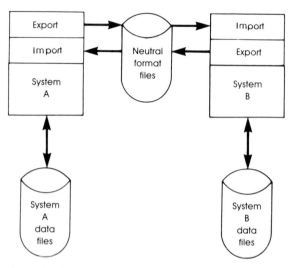

Data transfer via neutral files.

The development of these standards has not been without its own problems. Discrepancies are known to occur with data read into a system from a neutral format file, which has itself been created by the neutral format generator routine associated with the same system! If the use of a neutral format, for transferring data between different systems is a requirement of a company, then this 'back to back' test should be explored prior to purchase. This test should be followed by testing the transfer to and from sample-known destination systems.

The use of an internationally agreed neutral format for exchanging graphical data, has to some extent been subsumed by an emerging market dominance of a small number of draughting system vendors. These are aimed at general draughting, and to some degree, modelling market. Each of these now tend to support the data format of one of these systems, known as the DXF format. This has become an additional *de-facto* standard within many industries. It can be used not only for transferring drawing data between draughting systems which support it, but also to other applications such as computer based desk top publishing and word processing systems.

Retaining the accuracy and integrity of data when it is transferred between systems can cause problems. A classic example concerns the accuracy of complex surface representations using surface modelling techniques. It is important to ensure that each of the systems is capable of modelling such surfaces by using the same types of surface patches and handling the same orders of polynomial expressions which are implied. Failure to ensure such compatibility can result in small variations in surface definition. This can cause quality problems in the finished components, particularly where they need to match at surface boundaries.

Perceptions relating to the technology

Problems occur because of a number of perceptions which tend to abound within the management and operational staff of a user company. The responsibility for making a success of the introduction of computer aids must initially lie with the functional management of the units within which they are used. Much of the potential for success lies with the possibility for re-using data which has been input into the computer and subsequently processed within it. The operational management will not realise the full potential of the technology if they do not become fully familiar with the scope of the technology deployed. Too often they do not commit to learning the full potential of the systems which are at their disposal and consequently, they fail to plan for realising the full benefits. In most cases, additional benefit can be achieved, for example, by customising user menus with the tools available within general purpose systems. This customisation can help to relate the operation of the system to the work undertaken within the company. The technical management of the company need to understand these possibilities and actively plan for such customisation and to ensure staff are adequately trained in the methods needed. Other examples include failure to make use of facilities to store and process data fields which are of benefit in controlling the company and failure to output useful reports.

Resource problems

Similarly resourcing problems can arise as the technology becomes more widely used within a company. The availability of terminals, disk memory and computational resource needs to be kept under review. Problems of the system's operational performance can emerge quite suddenly if the system becomes overloaded in any of these aspects. The functional management need to liaise closely with the administrator or regular users of the system in order to plan to pre-empt resource limitation problems before they occur.

Senior management

Whilst much will depend on the middle levels of management, the commitment and perceptions of the senior management concerning the investment in technology aids are crucial. There must be a clear understanding of the full impact of introducing such technology amongst senior management. It is not sufficient merely for senior management to rely on the superficial perceptions about the benefits of computer based aids as are found in the technical press. They need to include the plans and operational consequences, implied in the deployment of such aids, within their overall strategy for the development of the company. They need to give their support to the technology and to maintain a top down pressure to ensure that the maximum benefits are realised.

Management training standards

There are particular problems associated with maximising the advantages from the technology within the UK. These are largely related to the low level of educational attainment and subsequent management training standards found in staff at the middle and senior management levels of many small and medium sized companies. This is particularly pronounced when compared with the standards found within much of Western Europe, North America and the Pacific basin. Many non-technical senior managers in the UK tend to have a low level of basic technical education and still less computer literacy. Many are hardly numerate at all. What is more, in the UK it is socially acceptable to proclaim a lack of numeracy as something of which to be proud rather than as a source of concern. The result is that too many senior managers almost rejoice in their lack of knowledge of the technology, or they express naïve and over enthusiastic perceptions and views about computers and how they can be used within their company. Too few senior managers in the UK have any reasonable level of awareness of the implications of deploying the technology for the other aspects of the business.

Capital investment

These problems are compounded within the UK by the dominant need to justify any capital investment often within very restricted criteria. A satisfactory financial return, based on simple direct cost savings over a short time horizon is frequently sought (often as short as two years or so). A broader perception and more strategic thinking is needed which takes full account of the changes required in planning the business, its products and methodology around the deployment of technology aids. This more thorough approach tends to be found in much of Western Europe, North America and particularly in Japan. It is still found too rarely within the UK. The UK has enjoyed a low labour cost economy compared with its major competitors and the pressures to invest in technology aids has, in part, been reduced as a consequence.

In many areas of high added value engineering, the capital investment cost in order to remain in the business often increases as the level of technology demanded increases. Advanced technology aids have become essential in many markets in order to produce products and systems of adequate quality, at an acceptable price, and within a lead time which is demanded by the market place. Many products now demand a level of quality and reliability which demand more rigorous design and optimisation than was the case in the past. To meet these growing challenges, computer aids form a part of the enabling technology in which investment must be made. Those companies and nations which have recognised this need and planned accordingly have increased and will continue to increase their market share, profitability and growth.

The alternative is for a company to become a producer of lower technology goods at a competitive cost. Developing nations are increasingly claiming these lower technology products as their own and have successfully increased their market share. For a company based in the UK to be successful in lower technology products, it would demand achieving lower unit costs than are likely to be found from the overseas competitor. For any level of material costs, investment has to be made to achieve much greater manufacturing throughput for a given level of labour input. This must be matched by aggressive volume marketing in order to compete on unit labour cost. The option of even lower labour costs for the UK seems an unlikely goal in the medium term future, although it may become the UK's ultimate destiny to become the Hong Kong of Europe! Increasing the throughput of lower technology products, with an attendant reduction of unit labour cost may be achieved with increased investment in manufacturing technology. In such a strategy, it would be necessary to take a longer term view of investment recovery time and to amortise capital cost over the higher volume of product and over this longer period. With this second option, much would depend on the company being highly successful in its ability to sell the product in high volumes. With regard to UK companies, this second option looks to be a difficult row to hoe, given the continuing general weakness in sales and marketing skills within most UK volume production companies.

There are difficulties in introducing technical computer aids within a company. Many of these problems relate to people; their skills, aptitudes and perceptions. They need to research, plan and train in order to maximise the potential benefits from investing in such technology. It is however prudent to note that, on balance, most companies find the advantages over the medium to long term investment outweigh the initial teething problems encountered.

Conclusion

These days, success will inevitably depend more on the people within the company and how the technology aids are deployed and used, rather than shortcomings in the technology itself. In order to take advantage of such technology the basic structures of good management of the company must be in place. The technology aids will not remove the responsibility of management to manage the company. They can only form an aid in the successful and proper management and control of the company. Managers must plan their company accordingly and ensure that they both extract and use the maximum amount of useful management and control information from the systems deployed. This demands that they become involved in planning the strategic use of the technology from a position of informed knowledge. They collectively need to know what is actually possible and feasible using the technology within their own companies.

Further, management need to assess the extent to which, if more information was available in a timely manner, more help could be derived from using the technology. They need however to be selective in the information which they pass to individual staff so as not to pollute their desks with data which will never be read or acted upon.

These are all skills which good managers have and which poor managers have not yet or may never develop. The urgency is to increase the skills amongst the management team to take best advantage of the technology. Managers who cannot acquire these skills can inhibit the development of the company and should be removed from the management structure.

14

Future horizons

The progress in providing usable computer aids for technical staff over the last 25 years has been remarkable. The technology started where aids were largely limited to numerical calculation based on engineering formulae carried out within the large high technology based companies and universities. It now covers a wide range of textual, numerical, graphical and topological design, and manufacturing activities within even the smallest companies. The initial investment needed to start to use the technology has fallen to a level where powerful computer aids can now be purchased for about the same cost as a typical manager's company car. This has largely been a result of the dramatic reduction of computer hardware costs and the volumes in which complex application software is now sold. The combination of these factors has resulted in affordable computing power becoming available to a wider range of companies.

As the market for such technology has grown, the scope has widened across the full range of high and low technology engineering. The developments in application software have been aimed both at providing general purpose solutions such as analysis routines, computer aided draughting/modelling and numerical control part programming and also in specialist niche markets. The specialist systems now penetrate such diverse applications as the design of kitchen layouts to the most erudite aspects of nuclear engineering. Often, such specialist and targeted systems have been built upon the basic operational software tools which have come from the development of the generalised systems.

The limitation of obsolescence

Clearly, over future years, further enhancements can be expected both in hardware and software techniques. It will continue to be the case that the vendor of any system will have new enhancements in preparation. To this extent the system which is bought today is already on the way to becoming somewhat out of date within a matter of a few years. Historically this has been more true in the context of the hardware components of the system. Application software develops perhaps more slowly, being highly labour intensive and following market feedback from its users. Notwithstanding these trends, if the application software runs under one of the widely used operating system environments, the purchaser of a system is, to a degree, protected against the pace of computer processor hardware development. This is due to the fact that any new computer processor which comes to the market, seen to be superior in respect of its price/performance, can only be taken up by the market if users can run application software on it. Otherwise, it will remain a fascinating tool of academic interest only.

Thus, any better value for money computer processor must support the widely used operating systems already present in the market place. Since the application software runs under the control of the operating system environment, the application software will run on any new processor which supports the same operating system environment.

Overall, the new or existing user company should not be intimidated against making a commitment to installing or expanding its computer aided systems because of the pace of development within the computer hardware (and related operating system). Developments will undoubtedly occur in both computer hardware and operating system environments over the coming years. But, the continued use of application software is governed more by its compatibility with its operating system environment than such factors as faster processors and developments in peripheral equipment. New hardware developments can be taken into account in reviewing hardware resources from time to time. Hardware elements of a system often can be upgraded by trading in hardware against new equipment; or possibly redeployed to new application areas as the technology develops within the company. It is perhaps prudent to base depreciation on the capital cost of all such hardware over a period of about five years in both cost justification and end of year accounting within the company.

Evolution of operating systems environments

From time to time, a new operating system environment will be introduced as a result of longer term research and development. These developments will be stimulated both by the development of new generic types of applications for computers, or in order to take advantage of new developments in computer hardware architecture or in the associated technology - such as data communications. Historical examples of these stimuli include the need to support multiple users on the same processor, the demand for handling larger application program modules, alternative user interfaces such as using on-screen windows and the demand for distributing computer power to the user community over data communication networks. Only when application software developers are convinced of the advantages of these new operating system developments will they commit to making their application software compatible with the new environment. The impact on the engineering application development tends to be a slower process and is usually provided as an upgrade path from the application software or turnkey system developer. This upgrade path provides the additional functionality supported within the new operating system environment within the context of the engineering application.

Development in peripheral devices

Just as processor hardware has evolved rapidly, so perhaps to a lesser degree has the technology of the peripheral electromechanical equipment such as disk storage units, printers, plotters and graphics scanners. Both the speed of operation and resolution of hard copy output devices, such as printers and plotters, and input devices such as scanners will undoubtedly continue to be improved. Improved performance of these devices will become available at lower capital cost. These developments will continue to be stimulated as designers take advantage of faster and less expensive electronic components. The use of new materials of greater strength to mass ratios will assist these new developments. Similarly, new methods of storing ever larger volumes of on-line data are being evolved. These methods will be based on optical disks using laser technology as well as better magnetic materials for storing information upon. Such developments will enhance the performance of the total hardware configuration.

Graphics terminals offering improved spatial and colour resolution for any given capital cost are continually being developed. Also, the functionality of such terminals, in terms of handling frequently used interactive graphical function, will continue to be enhanced. Such facilities as windowing, zooming, panning, dragging of graphical images in real time around the screen are serviced increasingly by hardware features built into the terminals. Such enhanced display terminal functionality is achieved by building graphics display processor integrated circuits, together with random access memory integrated circuits, into the terminal. Such enhanced functionality reduces the computational load on the computer processor on which the application software is supported, provided that the application software takes advantage of the hardware functions available within the terminal. Most graphics terminals provide such functionality and the scope is likely to improve over future years. The price/performance of graphics terminals has improved and is likely to continue to do so in the future but more in line with other peripheral devices.

User interface developments

The development of the user interface to computer systems has remained rather static over recent years since the advent of multiple active windows with hierarchically structured pull down and pop-up on-screen command menus. Application software developers continue to refine their menus with imaginative use of graphical icons and colour. Graphical icon techniques have also been used to feedback analysis results. Cryptic information to the user in such applications as circuit or thermal sensitivity analysis. There is clearly scope for such techniques to be applied in other areas. Similarly, in applications like logic or circuit simulation, selected output signals from the simulation are displayed in graphical form on the graphics display screen in a way similar to that shown on an oscilloscope with which the design engineer will be familiar. Even further, such display routines for simulation output signals can be controlled and manipulated in the same way that controls on an oscilloscope can be manipulated (e.g. varying the time base or trigger referencing).

This simple example demonstrates how sympathetic attempts can be made to liken the use of computer aids to using methods which are familiar to the engineer. Such improvements in the interface between the user and the computer application are likely to increase as the existing users place demands on the application software developers. Software writers will develop appropriate software tools and techniques incorporating them into their software routines. There is continued pressure to make the applications more compatible with the traditional tools of the trade. This is perhaps more easily achieved within applications aimed at niche markets where familiar methodology is more defined and less case dependent.

Within the engineering applications which are of a more general nature (for example general purpose computer aided draughting), there are signs that system developers are striving to develop more intuitive forms of user interfaces. The problem of the user having to interactively negotiate his way through the hierarchy of screen menus, is starting to be addressed. On-line context dependent help facilities are now virtually standard. However, such assistance is inferior to a more intuitive approach of indicating to the user what options are available at any particular time, depending on the mode being worked in on the system (e.g. creating geometry or editing geometry). In a few experimental systems, the system intuitively and automatically provides guidance to the user on options available at any state of command selection. The guidance is related to the position of the cursor in relation to the geometry which is active on the screen at any time. This approach allows options which have not previously been set up as default conditions by the user, to be selected in a more intuitive manner.

This more intuitive approach is a natural development from the existing methods of setting up options which will be invoked during the execution of system commands (e.g. forcing or 'snapping' a connection of new line to the end of an existing line, or snapping to the midpoint of the existing line). It provides a further example of how the problem of improving the interface between the user and the computer may develop over time. The choice of the user interface will always be a subjective matter. As was discussed earlier, different users prefer different methods of communicating with the system they wish to use.

Interpreting scanned data

Another area of research concerns the problem of capturing graphical information held on existing drawings which has not been produced by computer aided drawing methods. Optical scanners were discussed earlier and it was stated that these devices produced an image model within the computer which was not compatible with say conventional computer aided draughting systems. The reason for this incompatibility is that the image, after it has been scanned, consists of an array of picture elements (or pixels) to which a level of brightness (or grey level) and colour attributes (if appropriate) are assigned by the scanning device. Thus, the resulting scanned image within the computer does not model the lines, arcs, text etc which make up the drawing. It merely models a picture of the drawing.

This form of image can of course be held on a computer file and recalled and displayed on the user's screen. Further, the image can be modified by an authorised user of the system by means of interacting with the pixels of which it is composed, and changing their attributes or wiping them out altogether. Sections of the image can also be extracted and magnified, reduced or stretched by varying amounts in the horizontal and vertical direction. Software for other image processing techniques are also available which can perform such functions as filtering operations, data compression, smoothing or sharpening of transitions of brightness within the image or assigning false colours to the image. These software techniques have been developed in the context of processing other statistised images, typically within applications such as satellite data, medical scanner or robotic vision image processing.

Thus, high quality scanned graphical images can be captured, optimised and held within the computer. They can be useful in their own right in such applications as desktop publishing where such scanned images can be merged with text to produce technical documents of high quality. They can also find a place within information retrieval systems associated with catalogued information. For example a library of parts which is appropriate to a company's range of work can be held and maintained within a database. This component and sub-assembly database could not only contain all of the technical information about the parts, but also a statistised image of one or more views of the item. The image can be offered to the user as an on-line aid to recognising the part and can help in reducing the risk of a designer redesigning a similar part again. This is one use for the statistised image within a suite of computer aids. The company has to scan each of the catalogued items into the system and link this image to the data records associated with that part within the engineering database. It has to be asked if the cost incurred in setting up the information and maintaining it is going to be justified (particularly when the alternative is to simply refer to a printed catalogue entry for the image of the part). Again the choice will depend on the application.

Techniques are now in development to convert statistised images of drawings into the same structured forms as are held for drawings or models which have been generated within a computer aided draughting or modelling system. Consider what is required of the human brain to interpret such a statistised image of the drawing. The first stage in the process is to locate the various classes of geometric entity which are implied within the scanned drawing image. To the human eye the straight lines, arcs, circles, ellipses, smooth curves etc are self evident, even though they are made up of small dots of different levels of darkness, rather than consisting of continuous lines, circles and arcs. The human brain has no difficulty in interpreting the combinations of dots within the statistised image firstly into an engineering drawing of some general class, such as a mechanical detail drawing or a schematic logic diagram. Secondly, within this context the human brain recognises the combination of picture elements which clearly represent a straight line, circle or a character within a field of text. What is more, even a human brain which has not received any technical training can easily identify the sets of lines which combine to form the boundary drawing frame of the drawing. The drawing frame contains the administrative information such as the drawing number, its title, who drew it and checked it and the dates which apply.

Automatic interpretation

Consider the task of programming a computer to reliably simulate the same basic interpretation process - just for the case where it is told from the outset that the scanned image is that of a certain class of engineering drawing. The first task is for the computer to consider all of the pixels within the statistised image and to identify sharp transitions between adjacent sets of black and white pixels which represent lines or text on the original drawing. This assumes that some conventional image processing has been undertaken to identify and remove the coffee stains and other blemishes which frequently occur on an old original drawing. Stains will have been initially converted into irregular shaded areas on the raw statistised image.

Having got the computer to identify the intentional marks on the paper, it is necessary to identify which are drawing lines and which are associated with text. Subsequently the parameters concerned with particular drawing entities such as lines, arcs, circles or splined curves must be determined. For example, it would need to be able to identify a geometric entity like a circle with its centre and radius from the sets of pixels and so on. Further, it would need to identify the dimension lines and their associated text as being different from the geometry lines within the drawing. There are many sources of such potential ambiguities which would need to be resolved in order to produce an acceptable conversion for drawings of any general kind found in engineering.

In order to make these decisions the computer requires a considerable number of tests and rules which it would need to apply to the information extracted from the scanned statistised images. This process remains a major challenge and a topic of research. It is unlikely that a general purpose solution will be brought to the market in the near future. However, if the class of drawings is restricted in some way, in which the number of geometric combinations is limited to a well defined set of shapes, progress can be made. In particular, the conversion of such drawings as electrical or electronic symbolic schematic diagrams like switchgear or logic diagrams is meeting with some success. These diagrams are non-dimensional and the symbols in them can be used as patterns against which candidate data, extracted from the processed statistised image of the drawing, can be matched. Similarly electrical connections can be identified from the lines within the symbols. This subset of the general problem is a more constrained application area within which some success is now being reported.

Automated design

Throughout this book the emphasis has been on computer based systems which <u>aid</u> the user in various aspects of engineering work. In the majority of cases the computer does not automate the design, draughting or analysis process. Computer aids which are aimed at broad classes of general engineering applications must be limited to aiding the user, since the design methodology varies widely from application to application. This is particularly true within mechanical or civil engineering, not least because of the long history over which a wide range of methods of realising a design have evolved. Notable examples, where a degree of automated design using the computer has been economically achieved, include such cases as the automatic routing of printed circuit board tracks or automatic routing of pipework within process plant.

Other examples of automatic design systems are characterised by cases where the design process is closely regulated, often within a design code of practice. The design rules dictated by the code of practice can be built into application software and much of the design process can then be automated, once the overall design parameters for the design have been input into the system. Examples of such automated design include design systems for such products as heat exchangers, pressure vessels, cable sizing, portal frames and beams of various section. It is likely that this process of providing automated design systems within relatively narrow areas of design activity will continue.

More niche software products will appear on the market, particularly where design methodology is imposed by international standards and codes of practice. Nevertheless, the majority of systems will continue to be in the form of aids to design rather than automatic design tools.

Intelligent systems

The whole area of building so-called 'intelligence' into computer systems has received much attention in recent years. Such intelligent computer aids would have a potentially strong attraction to the engineering community since they could be used to maximise best practice in the engineering processes. In fact, the notion of building intelligent computer systems goes back over 30 years. A system operating with about a hundred rules was capable of interpreting and executing standard engineering calculus methods as early as 1961. Throughout the intervening years many researchers have got excited at the idea of combining ever more powerful computing technology with such technologies as image processing (based on raster images similar to television pictures) in order to build general purpose robotic systems. These were variously designed to move around their environment taking appropriate avoiding action, to inspect items for errors, locate items and orientate them in order to pick them up and place them within manufacturing systems.

Rule based technology within broad applications

Outside the engineering domain many attempts have been made to use sets of rules in conjunction with databases - for example to automate the translation of languages. It was hoped that such general purpose problems would yield to rule based systems, and all that would be required was to get the rules correct and provide the system with sufficient data on which the rules could interact. Many of these hopes, for the time being, have been seen to be over optimistic. The major reason for this failure relates to the difficulty in removing sources of ambiguity in both the rules which must be applied to such an approach, and ambiguity within the data as it presents itself to the so-called intelligent system.

To indicate the types of ambiguity problem which are encountered, reflect upon the following examples. Consider the case of using television technology combined with image processing software within a computer based AI system. This could be used for inspecting goods for gross defects, as the goods flow through an inspection point. In principal, if the outline geometry of the item is known within some standard form of computer model, the processed television frame image of the item to be inspected could be compared with the computer model. Any production items which had different images to the standard could be automatically rejected. The problems start to arise when the items to be inspected are presented to the inspection television camera at any orientation. The image of the item then has to be manipulated in order to identify a reference datum point on the object, and to transform the image further in order to seek a match with one of the standard orientations derived from the master computer model of the item. The computational load for such processing can be high, even for objects of modest complexity. This can soon result in problems because of the time taken to make the decision as to whether the item is good or not. If the process takes too long the rate of inspection must be reduced. This together with the capital cost of such a system, could result in the system becoming uneconomic. Similar problems of ambiguity can be found in pick and place robotic systems, where the task of recognising the orientation of the item to be handled can present problems. Much depends on the context in which the system's rules have to be applied and not least the quality and quantity of data which has to be tested against these rules.

In the case of the problem of generalised language translation, the ambiguities caused by context and usage can be demonstrated by the case of an attempt to translate such a phrase as 'out of sight, out of mind'. A computer translation may well result in the words 'invisible idiot'. This could be considered as an extreme case, but it does show the problems of applying rule based technology within wide applications and in which the interpretation of information is highly dependent on the context in which the information is being used. The broad range of engineering design, documentation and practice is one such general area. Thus, the prospect for general purpose AI systems within the near future is remote. Having said this, if narrower application frameworks are identified which can be represented within closely bounded areas of application, the potential for success is improved markedly.

Expert computer systems

In particular so-called 'expert' computer systems are being widely researched in an ever increasing range of engineering applications. The approach taken is to represent knowledge about a particular topic within a set of hierarchical knowledge models in the computer. The knowledge can be expressed as rules and declarations of properties about objects used in engineering information. The objects can themselves be structured in a hierarchical form and properties can be inherited from parent objects higher up the hierarchy of objects. For example, an object such as a particular transistor used within an electronics system may inherit properties from the general class of transistor type (e.g. its maximum operating temperature). These declarations and rules are designed to enable, for example, particular aspects of design or manufacture to be examined in the context of a proposed design data set.

By way of example, such a system could interrogate the computer held model of an electronics schematic diagram and its associated printed circuit board topological data in order to identify aspects of the design which would prove difficult to test. In this case, the expert system would need knowledge models of the the following topics which relate to the case of testability:

— The functional behaviour of components (truth tables etc);
— Connectivity information for the components (e.g. power connection pins, clock pulse input pins);
— Rules about good testability (access to test points or edge connector pins, safety considerations, isolation of signals);
— The functional capability of available test equipment (bandwidth constraints, test power supplies available, maximum clocking frequency, logic voltage levels available);
— Connectivity and topological models of the circuit design under tests.

In addition, rules need to be modelled which allow the interpretation of the context in which components are being combined in the circuit to be assessed, particularly for cases which have a significant impact on testability of the circuit. A typical example would be the combination of circuit components which imply the use of an on-board clock pulse generator within the circuit. Clearly in this case, the system would need to recognise the sub-circuit from the design data and to recommend that provision should be made in the design to isolate the on-board clock in order to inject clock pulse from the test equipment. Other such contextual patterns of components would need to be modelled together with their remedial recommendations.

The essence of such an expert system is that it can infer breaches of the hierarchical set of rules which cover the domain of the application, in this case testability, from the design data presented to it. Having identified the nature of the breaches, the expert system could prioritise the breaches of testability criteria it finds and present them to the designer, together with the following information:

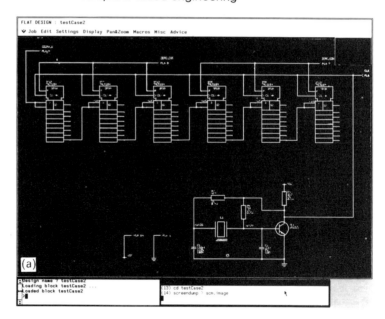

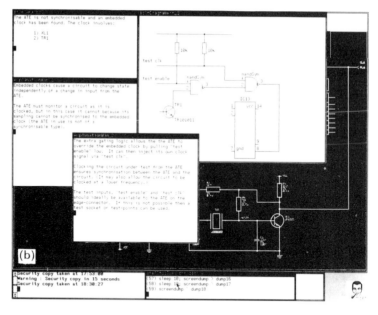

Prototype artificial intelligence system for advising electronics designers on the testability of circuits and printed circuit boards: a) Part of circuit schematic to be assessed; b) Identification of a testability problem, with explanation and advice on resolving the problem, as output by the testability advisor (courtesy of SD - Scicon plc and UK Ministry of Defence).

— An explanation why the proposed design is inadequate;
— The severity of the breach of good design practice;
— Suggestions of the remedial action which could be considered, relating to the context of the circuit components, connections between the components and the available test equipment (including identification of any spare circuit elements which could be used in correcting the problem e.g. spare gates);
— A warning of any other checks which need to be made as a consequence of remedying the problem (e.g. loading, fan out);
— An overall penalty assessment in terms of additional connections, components and surface area on the board which are required to optimise the testability of the whole circuit.

Provision must be made within such an expert system to acquire new knowledge. This new information will be in terms of clarification of, and extensions to, the testability rules and component property information. This ability for handling modifications ensure that the module's performance could be optimised - both over time, and in the context of its operation within a particular company design and test environment. Use of these 'learning' facilities would need to be carefully controlled and used by nominated staff charged with the responsibility of developing the testability criteria within the company.

This example of an expert system would be used in association with an existing computer aid for schematic draughting and printed circuit board design. It serves to demonstrate, in outline, one of the major development areas within the growing technology. The emphasis is on making analysis and guidance on achieving better practice in the design and manufacturing processes available to the wider engineering community. It is as if an expert engineer were sitting at the shoulder of the user giving advice and explanations to the user about problems which they may otherwise be building into their designs.

A number of such expert systems are now in the process of being developed. They cover a wide range of engineering application areas and are likely to enhance the benefits of computer aided methods within the foreseeable future. Often they will become add-on modules to existing, more general purpose computer aids. Many will be of interest to small and medium sized companies who may well lack specialist skills covering aspects of design which such expert systems are likely to address.

In parallel with these developments in application software, work is in progress with hardware architectures aimed at much faster processing of knowledge models. In particular, with the advent of low cost powerful computer processor integrated circuits, the way is now open for exploring such computing structures as neural networks. It is unlikely however that they will have an impact on the small and medium sized companies in the near future.

System integration

Much has been said and written about so-called 'islands of automation' by which is meant the separation of computer aids which individually address discrete engineering tasks. For example, separate application software modules have been adopted within many companies for tasks such as draughting, analysis, part programming and production control etc. Too often, both vendors and user companies have not paid sufficient attention to maximising the potential of the technology by passing the optimum amount of information between these different isolated application areas. The pressure for greater integration of computer aids has been strong for a number of years. Integration of computer aids within the technical functions of the company is clearly a prime objective. Also, duplication of administrative data held within the technical computer aids and within the commercial and accounting data processing should be avoided. Consequently, there is now pressure to forge greater integration between these two major classes of computer system.

One early school of thought suggested that each of these application areas should interact with a central pool of company information held within a common integrated company database. The more dominant approach has been for each major application area to communicate with its own database and to link these databases in such a manner as to avoid data duplication wherever possible. An example is the extraction of bills of materials and parts list for a project, or product, from a data model held in a computer aided design system, and to link this to a product structure database and part number master information held within a production planning and control system.

Progress towards greater integration between computer systems will offer greater potential for avoiding ambiguities within company data. As the supply market matures and the user pressure for links between computer systems increases, greater integration will be achieved and offered to the market. The consequential linking of the 'islands of automation' will result in improvements to overall information monitoring and control within all companies electing to use the linked technology.

Since in many instances, the modules which need to be linked together will originate from different developers of application software, there will be further pressure for improving the data exchange facilities discussed in the last chapter. The pressures for greater integration will no doubt result in further rationalisation of the supply market. Companies with software systems which increasingly need to work in closer harmony, will themselves either work more closely or merge with each other.

Educational trends

As the technologies outlined within this book continue to prove of greater benefit within the engineering community their use will grow, not least within the smaller and medium sized companies. The pressure for skilled and trained management and staff will grow, as it has in the last decade. Already this need has been recognised within the major higher educational establishments. Students are now exposed to the potential of computer aided techniques within their formal training. This training is still rather patchy and will need to develop further if the full potential of the technology is to be realised within industry.

Academia will undoubtedly respond to the demand for a greater awareness of computer aided methods amongst their students. There is however the more difficult problem associated with the level of awareness, and more importantly, competence in the control of this technology at the middle and senior levels of management within many companies. This problem is probably at its most severe within the small and medium sized companies. The day to day pressures at these levels of management are great and the time for exploring new methods and learning new skills is limited. These pressures are not helped by the need to re-assess many of the fundamental methods of engineering practice in order to relate them to the new technologies.

If the full benefits from the technology are to be achieved, the management of companies must consider the planning and control of the technology as an integral part of their overall strategy for the company. They must equip themselves for this task. After all, it is not the technology which is in itself important. When a company invests in the technology, the greatest impact will be brought about by the managers who plan and control its use and by the skills of those who come to routinely use it.

Index

Illustrations are referenced in italic

access,
  unauthorised, 134
accounting, 5
AI, *see* artificial intelligence
alpha-numeric data, 5, 56, 77, 79, 162
application program, 6, 7
application software, 6, 7, 8, 14, 28–9,
  56, 74, 134, 135, 205, 206, 214,
  238, 251
  modules, 11, 79–80, 210, 223, 245,
    280
  suites, 7
application software tools, 71–9, *80*,
  270, *see also* individual tools *e.g.*
  language processors
archiving data, 134, 231–32, 254
artificial intelligence, 29, 50, 52, 77,
  116, 143, 146, 275–79
assembly, 124
assessment team, 209, 211, 212, 215

associative,
  dimensioning, 130
  hatching, 130, 212
audit trail, 88, 178
auditability, 1
automated design, 274
automatic,
  cross hatching, 130
  routing, 7, 11, 274
  testing, 77, 88

background mode, 4
back-up practices, 227
  full, 228–9
  minimum, 229–31
bar codes, 190, 192
batch operations, 1
benchmark testing, 209–12, 214, 215
  final system selection through, 210

best practice, 139, 175
 design, 144, 279
 manufacture, 144, 279
bills of materials, 169–71, 172, 174, 224
Boolean logic, 36, 99

capital cost, 207, 270
capital investment, 19, 263
CAPP, *see* computer aided process planning
central processor, 28, 56–7, *see also* hardware
 disadvantages of, 57
 advantages of, 58
chemical process simulation, 7, 11, 12, 37, 106 *see also* computational load
chip topology design tools, 77
circuit diagrams, 11, 39, 76, 119, 124
circuit sensitivity analysis, 270
clamping fixtures, 156
clock rates, 4, 11
codes of practice, 24, 25, 37, 79, 122
colour information, 3, 8, 59, 62, 86, 94, 103, 107
command processor, 73
commands,
 alpha-numeric, 10
commercial data processing, 1, 6, 242
communication, 200
 between design and production planning functions, 137–38
 between engineering users and computer aids, 10
company,
 data, 92, 225, 252
 functions, 84
 policy objectives, 159
 priorities, 160–62
 standardisation, 90
 standards, *see* standards
 systems, 257–59
 technical review of, 199–203
components, 124
 alignment of, 114, 121
 feature, 146
 information, 9, 10, 113
 libraries, 133, 212

components—*cont.*
 milled, 142
 obsolete, 171
 parts list, 77
 reference, 171
 standard, 120
 technology, 249
computation speed, 2, 8
computational load, 7, 126, 270
computer,
 graphics, 2, 12, 13, 152
 hardware, 14, 27, 56–70, 126, 204, 207, 236, *see also under* individual hardware *e.g.* engineering workstations
 hardware costs, 267
 libraries, 13, 22, 90, 119–20, 142, 223, *see also* component, drawing and data libraries
 main frame, 4, 12
 memory, 2, 8, 71, *see also* virtual and random access memory
 mini, 4, 5, 7, 12, 13
 networks, 7, 9, 66–70, 75, 105, 134
 personal, 11, 13, 61, 118
 processors, 7, 9, 11, 28, 59, 237
 programmers, 132
 software, 28, 70–9, *see also under* individual software components *e.g.* operating system software
computer aided,
 analysis, 6
 animation, 25
 design, 10, Ch 6, 152, 157, 202
 in the fashion industry, 106–109
 draughting, 6, 12, 58, 106, Ch 7, 152, 157, 199, 202, 211, 271
 engineering,
 benefits of, 81, 82–5
 capital cost of, 92
 financial viability of, 84, 88
 initial assessment of, 82–3
 justification of, Ch 5
 prioritising problems, 84
 process planning, Ch 8
 generative, 145–48
 generic, 144
 language, 149

computer aided—*cont.*
  process planning—*cont.*
    staff in, 139
    variant, 144–45
computer aids,
  expanding the use of, 245
  problems areas, Ch 13
  reasons for poor results, 245–46
computing priorities, 58
conceptual,
  design, 10, 34–6, 93, Ch 6
    performance, 104–105
  modelling, 34–6
connectivity schedules, 128, 132
consultants, external, 83, 198–99, 215
contract of supply, 209, 215–16
credit notes, 182
cross-sectional profiles, 102
cursor, 10
customer credit, 181–83, 184
customisation of computer aids, 13, 27,
    110, 216–17, 257
cutting tools, 156

data,
  communication, 75
  compatibility, 12, 258
  corruption, 227, 232, 238
  exchange, 280
  files, 9
  libraries, 79
  loss, 227, 252
  master –, 9, 69
  processing, 30
  project –, 9
  networking, 60
  on-screen, 183
  recovery, 227
  security, 134, 238
  storage model, 79
  system tools, 72
  text –, 9
  transfer, 136
    software tools, 75
  volumes, 237
databases, 72, 133, 141, 144, 147, 152,
    162, 253, 258, 275

decision trees, 146
  action states, 146
dedicated link, 259
design,
  alternatives, 22–3, 27, 86, 94, 104
  analysis, 2, 12, 36–8, 85, 88, 216,
    223
  constraints between countries, 24
  cost performance of, 25
  costs, 88
  data, 141, 142, 145
  drawings, 38–9, 83, 97
  of external appearance of product, 94
  functions, 201
  houses, 103
  language, 120
  quality, 88, 105
  simulation, 2, 36–8, 88, 104
  testing, 24
  using features, 142
designer, 6, 42, 69, 96
detail drawings, 19, 38, 94, 97, 120,
    121, 201, 202
development,
  cycle, 87, 88
  timescales, 26–7, 86
dimensioning, 122, 129, 157, *see also*
    associative dimensioning
  chain, 123
  techniques for, 122–23
disk,
  access, 7
  storage, 4, 5, 8, 9, 11, 57, 68, 70
documentation, 39–40, 86, 89, 211
  as-built, 52–3
  factory, 189–92
  shop floor, 175, 187
  system, 210
  user, 210
domain expert, 79
draughting,
  parametric, 120
draughtsman, 6, 120, 123, 124, 131,
    132
drawing,
  amendments to, 127
  arrangement, 202
  characteristics of, 126, 127

drawing—*cont.*
  conventions, 123
  embellishments, 124
  libraries, 134
  modification, 129
  parametric, 120
  registry, 134
  retrieval, 134
drawing office, 19, 38, 66, 84, 126, 137,
    141, 223
  applicability of computer aids in, 118
  productivity in, 120, 126, 127–28
dynamic drag, 63

electrical schematic draughting, 5
electronic system simulation, 11
engineering,
  applications of computer aids to, 2, 5,
    6, 12, 136, 269, 276
  civil, 13
  companies,
    legal pressures on, 23–5
  data, 8, 58
  drawings, 2, 3, 10, 96, 127, 202, 273,
    see also drawing
    graphs, 4
  labour costs in, 19, 82, 85, 176, 263
  language, 73
  models for, 3
  quality considerations in, 20–2
  user market, 6
  workstations, 8, 9, 12, 13, 59–60,
    220, 236
equipment,
  malfunction, 228
  siting, 219–21
  electrical considerations, 220–21
executable code, 28, 29, 217
expert computer systems, 276–79

fabrication, 144
factory documentation, 47
families of parts, 144
feature,
  achievement, 147
  codes, 149

feature—*cont.*
  definition,
    operator assisted, 141–42
  recognition, 79, 142–43
features,
  generic, 142
  manufacturing, 142
  physical, 142
file,
  directory, 134
  handling, 6
finite element analysis, 2, 11, 37, 58,
    90, *91, see also* computational load
fixings and fastenings, 119, 172
functional performance, 103, 252
functional specification, 34, 128, 203–
    204, 209, 210, 212

geometrical constructions, 121
  replication of, 120
geometry, 115
  bounded, 151, 152
  calculation, 121
  construction, 121
  elements of, 128
  manipulation, 129
  master surface, 115
  move, 129
  parametric, 131, 132, 213
  structures, 142
  tool tip, 154
  transitional, 121
  variables in, 131
graphical,
  editing, 127
  image, 2, 3, 4
graphics,
  display terminals, 8, 28, 56, 59, 61–3,
    269
    colour range of, 62
    refresh raster, 8, 61
    refresh vector, 3, 4
    screen size, 61
    spatial resolution of, 62
    storage tube, 5
  handling routine tools, 73
  screen, 10, 107, 131, 154

graphics—*cont.*
  simulation, 156
  tablet, 10, 118
  windowing techniques, 123
grinding, 144
group technology code, 144

hard copy, 3
hardware,
  cost, 9
  problems, 243, 253–54
  upgrades, 245
  vendors, 206
hatching and shading, 123, 124
heat exchangers, 37, 110, *111–113*, 274
highway design, 5, 63

image processing, 50, 53, 272
incremental,
  back-up, 229, 230
  dump, 230
industry,
  aerospace, 23, 24, 37, 88, 101, 114–15
  automotive, 18, 23, 88, 101, 114
  chemical process, 51, 52, 99
  electronic, 2, 10, 37, 52
  fashion, 26, 34, 101, 106–109
  heavy electrical, 37
  manufacturing, 14, 19, 26, 34, 139
  mechanical, 14
  nuclear, 24, 37, 51
information,
  and resource sharing, 105–106
  transfer, 71
inherited properties, 276
input devices, 6
inspection, 49–51, 121
  drawings, 10, 39
  equipment, 146
  image processing in, 50
  printed circuit boards in, 50
installation of computer aids, 48–9
integrated circuits, 28

job,
  costing, 192, 194
  routing card, 189
jobbing shops, 151

knowledge,
  based systems, 77, 79, 116
  models, 276
  modelling tools, 77

labour costs, *see* engineering
language processor tools, 28, 73
layering, 130
lead times, 47, 48, 167, 172, 173
  production, 185
  shortening of, 85–6
learning curve, 234, 235, 252
local area network, 67, *see also* computer networks
logic,
  diagrams, 38, 39, 42, 76, 77, 119, 273, 274
  simulation, 11, 13, 37

machine,
  capability, 149
  tools, 5, 10, 46, 141, 146, 147
  numerical control, 150
  vision, 143
machined parts, 139
maintenance of products, 51–2
man-machine interfaces, 10, 79
manufacturing, 121, 124, 139
  logic, 149
  methods of, 138
market feedback, 135
master,
  data, 9
  file server node, 68
material,
  costs, 192
  management, 47–8, 169, 182
  requirement planning, 171, 173
  scheduling, 192
maths co-processor, 118
  chip, 11

mechanical,
    components, 25
    detail drawing, 5, 12, 62
    engineering, 144
menu icons, 29, 74, 270
menus, 29, 74
metal cutting, 147
middle management attitudes, 240–42,
    252, 281
milled pockets, 151
milling, 144, 154
minimum hardware configuration, 205
model maker's thumb, 121
modelling, 61
    complex double curved surface, 76,
        100–102, 109, 126
    full solid, *76*, *94*, 98–9, 102, 126, 141
    numerical modelling, 90, 115
    three dimensional, 2, 11, 28, 34, 94,
        103, 142, 212, 257
    tools, 76
    wireframe, 76, 97–8, *102*, 126
models,
    surface properties of, 103, 107
modems, 13, 70, 213, 238, 243
modular software tools, 28
monitoring of,
    quality, 20
    response times, 237
    software, 238–39
    staff, 233
    system, Ch 12
        administration, 242
        supplier, 243–44
moving mechanisms, 25
MRP, *see* material requirement planning
multi-sheet diagrams, 76

national standards, 24
NC part programming, *see* numerical
    control part programming
network systems, *see* computer networks
neutral,
    interface formats, 12, 259–61
    standard format, 75, 136
numeric resolution, 8
numerical,
    accuracy, 150

numerical—*cont.*
    control part programming, 5, 45,
        150–51, 212
    modelling, *see* modelling
numerically controlled machine tools,
    115, 157

operational,
    errors, 238
    loads, 203
operating,
    parameters, 146, 148
    system, 6, 134
        environments, 269–70
        licence fees, 208
        multi-tasking, 7, 11
        software, 70–1
output devices, 6

Palto Alto laboratories, 9
pan facilities, 131, 156
parallel running, *see* phased
        implementation of system
parametric draughting, *see* draughting
part,
    numbering, 162–63, 166, 169, 223
    programming, 154
        function, 151
        parametric functions, 131
        parametric features, 141
        software, 151
        system, 107, 152, 151–57
    programs, 150, 151, 201
parts,
    and materials lists, 39, 42, 46, 162
    and materials management, 164–65
    requisitions, 191
password,
    control, 69, 182, 185
    protection, 105, 134, 167, 170, 184
PCB, *see* printed circuit boards
perceptions of,
    management, 261
    operational staff, 261
peripheral devices, 6, 269
'permanent stop', 182
petrochemical plants, 20

phased implementation of system,
216–17, Ch 11
    parallel running, 226–27, 235
photogrametory, 53
picking lists, *174*, 175, 176
piping and instrumentation diagrams, 5,
119, 274
piping isometrics, 5
plot queue, 4, 133
plot spooling, 133
plotters, 5, 7, 61, 64–6, 211, 220, 253
    electrostatic, 65
    pen, 65
    photo, 66
plotting, 133
pop-up menus, 64, 74, 256
post processor, 156–57
power analysis, 104
pressure vessels, 110, 274
printed circuit boards, 5, 7, 11, *40–41*,
50–1, 58, 62, 77, 274, 279
printers, 61, 211, 220, 253
process,
    planning, 10, 42–6, 201, *see also*
        computer aided process planning
        manual, *44*, 45
    plant, 10, 48, 88, *99*, *100*, 274
    simulation, 5
processing speed, 8
processor memory, 4
product,
    life cycles, 25, 26
    physical topology of, 104
    ranges of, 89, 120
    surface texture properties of, 94
product/system life cycles, 25–6,
117–18
production,
    control, 46–7, 186–89
    department, 83
    detailing, 10
    engineer, 6
    management, 185–95
    planning, 19, 42, 46, 189, 199
        priority of objectives, 46
        strategies in, 46
    schedule, 172
    staff, 138
    time, 148

project management, 48
prototypes, 23, 87, 93
pull-down menus, 64, 256
purchase order processing, 48, 167,
177–80

quality controls, 21
quality of finish, 139, 144, 147
quality of information, 83, 123

random access memory, 5, 7, 8, 28, 59,
61, 270
raster,
    frame, 3
    screen, 8
    technology, 8
refresh raster graphics, *see* graphics
refresh vector graphics, *see* graphics
relational database system, 72
removable disk, 5
resource problems, 261–2
rework, 46, 88, 104, 105, 187

sales,
    'hit rate', 87
    invoices, 181, 184
    order processing, 180–85
    orders, 171, 172, 184
scanned,
    data, 271–72
    graphics, 75
schematic diagrams, 125
scrap views, 40, 49, 88, 97
screen,
    image, 4
    menus, 9, 64, 74
    repaint times, 136
    size, 4, 61
    windowing techniques, 9–10, 40, 74,
    256
scripts, 210–12
sectional views, 125, 126
senior management, 197, 199, 203,
234, 241, 262–63, 281
sequencing rules, 146

setting up time, 140, 192
shading, *see* hatching and shading
shoe industry, 26, 34, 36, 101, 106–109
shop floor data collection, 191
single processor systems, 6
skill shortages, 17, 18–19, 87
    lack of training in UK, 18
software,
    developments, *see* computer, software
    houses, 6, 104
    problems, 244, 254
    vendors, 205
solid model, 63
spatial resolution, 3, 4, 59, 61, 62, 65,
    133, 269
spread-sheets, 133
staff training, 222, 231, 234, 281
standards,
    company, 9, 22–3, 90
    international, 259, 274
    management training, 262–3
    national, 24, 25, 122
    quality, 139
stock,
    control, 47, 48, 165–69
        module, 174
    costs, 173
    movements, 165
    reports, *168*, 169
    status, 165, 166
storage tubes, 2, 3
strength analysis, 104
stress analysis, 12, 37, 106
sub-assemblies, 22, 36, 85, 89, 120
    compatibility with major product, 36
sub-assembly drawings, 201
supply market,
    for computer aids, 12–15, 250–51
    volatility, 250
surface,
    finish, 146
    patches, 101
switchgear diagrams, 274
symbolic schematic diagrams, 76
symmetry, 121–22, 129
syntax, 73
system,
    administration, 70, 134, 225–26, 235,
        238, 241, 256

system—*cont.*
    back-up, 227–31
    block diagrams, 76
    capital cost of, 135
    demonstration, 212
    developers, 7
    early use of, 221
    house-keeping, 70, 203
    failure, 135
    implementation, 216–17, Ch 11
    integration, 280
    monitoring, Ch 12
    response times, 5, 6, 7, 58, 135–36
    vendors, 10

tablet menu, 10, 64
tablets, 63–4
tape unit, 5
technical,
    interviews, 199
    management, 144, 233
    review of the company, 199
technology,
    antagonists, 239–40
    champions, 239–40
templates, 39, 64, 74, 119, 120, 121
'temporary stop', 182
test specifications, 201
testability rules, 277, 279
testing, 49–51, 118, 121
text,
    characters, 4
    data, 9
textile design, 63
textual information, 124–25
texture, 103, 123
thermal analysis, 25, 37, 104, 270
three dimensional modelling, *see*
    modelling
time sharing, 4, 5, 12
tool libraries, 22, 154
trade union movement,
    attitudes to CAE, 29–31
training arrangements, 213
translation program, 12
trial kitting list, 174, 189, 191
turned parts, 121, 142, 151
turning, 144, 154

turnkey system, 1, 14, Ch 10
  benefits of, 198
  vendors, 13, 104, 206–209

UK,
  drawing conventions, 129
  drawing office and production
      interface, 137
  labour costs, 19
  updating software, 254
USA, 1
  drawing conventions, 129
  market dominance, 14
  skill levels, 19
user,
  commands, 9
  community, 134
  directories, 231
  groups, 135
  interface developments, 270–71
  terminals, 7

VDU, 4
vendor,
  response, 205–209, 244
  selection, 209–14, 215
  support, 135, *see also* system vendors
vibration analysis, 12, 37

viewing commands, 154, *see also* zoom
  and pan
virtual memory, 4, 7
visual display unit, *see* VDU
visualisation techniques, 103

windows,
  facility, 131
  multiple, 9, 131
windowing routines, 28, 270
word processors, 133
work centres, 148, 150, 160, 173, 186,
  188, 193, 195, 201
  supervisor, 190
work in progress, 46, 173, 186, 187,
  192, 193, 227
working capital, 90
workload, 66, 201
works order, 171
  processing, 173–76, 176
  schedules, 172
workstation node, 9
workstations, *see* engineering

zoom facilities, 63, 123, 125, 131, 156,
  270